Design through Digital Interaction

Computing Communications and Collaboration on Design

Chengzhi Peng

intellect™
Bristol, U.K.
Portland OR, USA

Paperback Edition First Published in Great Britain in 2003 by
Intellect Books, PO Box 862, Bristol BS99 1DE, UK

Paperback Edition First Published in USA in 2003 by
Intellect Books, ISBS, 5804 N.E. Hassalo St, Portland, Oregon 97213-3644, USA

Hardback Edition Published in Great Britain in 2001 by Intellect Books, Bristol, UK
Hardback Edition Published in USA in 2001 by Intellect Books, Portland, OR, USA

Consulting Editor: Masoud Yazdani
Cover Design: Sam Robinson
Copy Editor: Holly Spradling

A catalogue record for this book is available from the British Library

ISBN 1-84150-844-6

Printed and bound in Great Britain by 4edge Ltd, Hockley. www.4edge.co.uk

Contents

Figures

Tables

Acknowledgements

The fact that I am the sole author of a book about collaborative design seems not to contribute much to the credibility of the authorship. However, the production of this book to me has been a long journey commencing many years ago when I was a research student at the University of Edinburgh. Through various points of the journey, I met people who shared similar interests in the subject matter and many of the discussions I had with them not only helped to shape my work at that time but also influenced my later view of how further work could be developed.

I wish to thank John R. Lee and Aart Bijl at the Edinburgh Computer Aided Architectural Design (EdCAAD) research unit, Department of Architecture, University of Edinburgh. As my Ph.D. advisors, their often stimulating questions and discussions seem to have kept my interests in the research area alive till today. I was fortunate to be given the opportunity to join the Computer Supported Collaborative Design project at Information Technology Research Institute (ITRI), University of Brighton. Working with Donia Scott, Lyn Pemberton, John Downie, and Simon Shurvile, I had the opportunities to have my work on collaborative design questioned by colleagues working in different disciplines. Since joining the School of Architecture at the University of Sheffield in 1995, I have been benefiting from being a member the Process Research Group led by Bryan Lawson. Through the Group's funded research, I was able to gain contacts with real-world design practice and software development.

Apart from dedicated project funds, my research over the years has received financial supports from the CVCP Overseas Research Student Award Scheme, the Sheffield University Research Fund, and an Award to Newly Appointed Science and Engineering Lecturers from the Nuffield Foundation. These supports have enabled my continuous working on this book in one way or another.

Finally, a warm thanks to my wife, Wanchen, for her constant encouragement and support, which has made my completion of this book less a struggle.

To My Parents

Preface

This book grows from a collection of my writings on topics related to collaborative architectural design processes and computer supported collaborative design. Computer supported collaborative design is a rich subject area that covers many component areas which by themselves are established domains of studies. To discuss some of the generic issues in this complex area, as intended by this volume, many connections need to be made. First and foremost, there is the understanding of how designers think and work together as distributed groups or centralised teams. Where is the evidence of teamwork and how can we describe adequately the nature and characteristics of collaborative architectural design processes? Secondly, academic and software industry groups have in the past decade developed many experimental and commercial computing systems aimed at supporting design collaboration. What are the ideas and concepts behind these system designs and implementations? Do they share similar agendas for system development? Then, there is always the latest invention in computing technologies that may open up new opportunities of harnessing computing communications for design collaboration. But how will new computing communications shape the ways that designers collaborate with each other? Or, should design studies of creative teamwork in design have a say regarding what computing technologies are worth developing?

I intend to bring observations, discussions and speculations on these issues into a single coherent volume. In many places, no connections among parts of the content are made explicitly. It is my aspiration that readers of this book will return to some sections from time to time and make that connection, relating their own experiences and interests in learning/practicing design, using collaborative design tools, or developing new features of teamwork-supporting systems.

With first reported applications of computers in building design practice in the early 1960s, we now begin to see the impacts of computer applications on design practice and education. In many aspects, the richness, complexity, diversity, and above all, the creativity in architectural design seem to have defied computerisation. Still, we often perceive contrasting attitudes and views about computers in architectural design. On one hand, the computers are considered as promising tools for working out solutions to contemporary building design and construction problems; on the other hand, digital tools and medium still appear alien to some people who believe that the sense of drawing and model-making by hands and eyes in direct contact with conventional media cannot be recreated or replaced in a digital way. Given that these fundamental differences will not be resolved in the near future, it is clear that making computers to support not only individual but collaborative design processes is bound to be even more problematic. A great challenge for system development in support of collaborative design is to accommodate both group dynamics and process management; the two ends are not always within the reach of a system's boundary.

1 From CAD to CSCD: An Introduction

Seen from the dawn of the twenty-first century, the use of computers in architectural design can be traced back at least three decades ago. What we see from the past is that many earlier predictions or expectations of how computers might fundamentally change the ways we design buildings, or more broadly, the built environments, did not come to fruition for various reasons. However, we can see the inventions of information technologies and their applications in the field of architectural design have been developed in rather diverse if not too fragmented ways. There has been a great deal of interaction between designs of software that aim to support aspects of design processes and actual uses of the software in design practice. As design software have become ever more sophisticated and complex, it may be contentious or even misleading to propose any generalisation of how we have arrived at the current position with regard to the development and utilisation of digital tools and media in architectural design. The starting point of this book was indeed based on a general observation that we have clearly evolved from Computer Aided Design (CAD) to Computer Supported Collaborative Design (CSCD).

Nowadays, both developers and users of design software commonly express greater interests and demands of more versatilities in computer networking while tapping into increased powers of data processing. The opening chapter of this book intends to present a view of the evolution of software development for architectural design, which may be sufficient to set up a background for latter chapters on the themes of computer mediated communications and collaborative architectural design. We shall follow an almost chronological order to review some of the major milestones of technological and theoretical development that have some long-lasting influences on the current state of architectural design computing.

In different contexts, the term "architectural design" can point to different domains of reference, namely design as a product or design as a process. In the real world, the distinction between product and process may be not that significant. To design practitioners, talking about a design process without referring to some specific products of design (i.e., sketches, drawings, models, or the actual buildings that have been built) can sound all very hypothetical. On the other hand, in academic research, generic descriptions of design cannot be achieved without looking at how designers develop design. These abstract descriptions of design processes may be not useful in learning how to do architectural design, but clearly the development of computer-based design environment has close connections with some of the descriptive models of design. Depending on which view of design and system strategy is adopted, computer-aided design systems can be either *product-structured* or *process-oriented*.

Another equally important aspect of how architectural design can be described is from the perspective of *human communication*. Although most famous buildings are reported and consequently perceived as achievements of some well-known individual designers or architects, we should be reminded that building design in reality is essentially group work; professionally speaking, it involves communication and collaboration of different designers playing different roles and bringing in different domains of specialism.[1] Contemporary building design practice displays even more vividly the necessity of communication and coordination among members of a design team. As demands from clients and changes in building technologies evolve, a single person is no longer in a position of attending to every facet of a building design project that may range widely from composing architectural expressions to specifying technological systems. The evidence of human communication can be traced by observing design as a product as well as a process. As it will be shown in later chapters, the recent progress and availability of computer networking has prompted CAD system development toward a new direction of "groupware," that is, CAD software developed to support a wide range of group design activities, such as design information sharing, collaborative design decision-making, virtual design conferencing, on-line product data access, and so on.

This book intends to bring together topics related to computer communications and design collaboration in a single coherent volume. Given the ever-accelerating pace of technological development in computing and communications, this book is not intended to unveil the latest technological systems available for real world applications but to discuss some of the important issues of Computer Supported Collaborative Design (CSCD). As a result from my own research, the resources and discussion in the book are organised into four main topics: (1) the earlier experiments in system developments of computing communications for supporting collaborative design work, (2) the features of team working in architectural design and the requirements for computer support, (3) new possibilities and opportunities for collaborative architectural design as opened up by computing communications, (4) the issues and challenges of developing future CAD environments as *teamware* or *groupware*.

The current volume was written with the following readership in mind: students and tutors of design computing in architecture who are interested in knowing the recent developments of groupware in relation to team working processes in architectural design; secondly, CAD systems researchers and developers concerned with the representation and communication requirements of team working in architectural design in order to deliver responsive computer-based group design supporting tools; and thirdly, architectural professionals who are contemplating the technological capabilities of teamware in enhancing group practice in the design and construction of the built environment.

1.1 Some milestones in the development of architectural CAD (1960s—1990s)

The research and development of computer-aided design in architecture can be traced back to the early 60s at least. If we look back across the decades, a number of stages can be identified as some notable milestones in the development of computer aided architectural design. A look at these historical developments may serve as a foreground ground to introduce the later developments of computer supported collaborative design. As we shall see, some of these historical system developments were end results of academic research, while others have been further taken up by industrial system developers and vendors.

Pioneering Integrated Systems. The application of computing systems to architectural design practice can be traced at least three decades ago. In the late 70s, some government institutes employed professional computer programmers to design and implement large mainframe-based integrated CAAD systems. These specialised dedicated systems were intended to embrace nearly all aspects of building design for computer-based productions of design drawings, specifications, and other required information for construction. Classical examples are the OXSYS system (Hoskins 1977), developed at Oxford for hospital design, and the SSHA system developed at Edinburgh for housing design (Bijl, Renshaw et al. 1970; Bijl, Stone et al. 1979). Notably, these early pioneering integrated CAAD systems were developed under two conditions: firstly, these integrated building design systems were targeted at particular regularised building construction methods (e.g., the "Oxford Method" of construction for hospitals in the OXSYS system, and the Scottish Special Housing Association Standards in the SSHA system); secondly, what contemporary computing systems could provide were highly centralised computing services, as pointed out by Bijl later in (Bijl 1989, 97), which could only be affordable and accessible to some larger governmental organisations or research institutes.

Performance Simulation and Analysis Systems. To building scientists and engineers, a building is a physical system whose behaviour or performance can be measured and predicted according to some mathematical models. One of the key areas of environmental performance in buildings is energy efficiency. Thanks to building scientists' efforts of drawing all factors relevant to the thermal behaviour of matter into computational models, it is possible to tell in advance how a building will perform environmentally prior to its construction. Another major area of interest is the simulation of lighting design in the built environment. Again, complex mathematical models have been established to measure and predict effects of natural and artificial lighting. The rise of computer-based performance simulation systems in the early 1980s came around about the same time as high performance workstation platforms running mainly the UNIX operating system was made more available. In comparison, this type of computing environment was more supportive of system development and more affordable than those of earlier mainframe systems. Dedicated research units in the UK, for instance, the ABACUS Unit at the University of Strathclyde, developed computer programs that aimed to deliver computer simulation of environmental concerns in building design such as energy, lighting, and air flow. The ESP (Environment System Performance) system developed at Strathclyde (Clarke 1985) is a classical example of a building performance simulation system in which a fair amount of mathematical modelling of building physics and mechanics has been achieved and verified.

"Dumb" Drawing Systems. Aart Bijl first used the term "dumb systems" to refer to the development of computer-based drawing systems in the late 1970s and 1980s (Bijl 1989, 137). Bijl gives an example of a "dumb" system—word processors. In a word processor, as Bijl puts it, words can be represented by an environment of characters that has no functionality of interpreting what can or cannot be words. By the same token, a dumb drawing system presents an environment of drawing/graphic primitives (such as line) plus a set of manipulative operations (such as rotating) in which drawings can be constructed and changed. Depending on the range of the basic graphic constructs a system may provide, users of a dumb drawing system can produce and manipulate drawings, which the system has no knowledge or models of. The meanings or semantics of drawings are, so to speak, outside the drawing environment.

Knowledge Base Systems. When less centralised and lower-cost computing systems were made available by the capacity of the IT industry achieved in the 80s, knowledge-based approaches to CAAD system engineering came into existence. This new trend developed novel techniques by combining several diagnostic and generative Expert Systems into an integrated modelling environment. Researchers at Carnegie Mellon University, among others, developed prototypes of knowledge-based design environment. This includes the use of knowledge-based programming techniques for interfacing knowledge base systems with database management systems (KADBASE, Rehak 1985), and the Integrated Building Design Environment (IBDE, Schmitt 1990). Being strongly influenced by Artificial Intelligence (AI), these systems aimed to automate some aspects of the design tasks by compiling sets of evaluative and/or generative procedures into computerised knowledge bases. An important concept of knowledge-based design systems is the "design space" which is defined by the set of all possible solutions of a design problem as computed by the system. The system then searches the design space for plausible or optimal instances by applying the rules coded in the knowledge bases.

It is worthwhile to point out the contrast between the graphics in a knowledge base system and that of a dumb drawing system. All graphical instances in a knowledge base system have meanings that are prescribed by the constructs and operations of the system. In a way, the drawing shown to the user is basically a kind of interface for inputting or manipulating data values as defined in the system's data structures. Unlike the dumb drawing systems, a user using a knowledge-based design system for the benefits of knowledge processing of design schemes can only draw things that are pre-defined by the system. For example, to utilize a system's capability of performing building construction cost analysis given a set of related 2D drawings, designers will have to work with the types of drawing elements and operations as specified by the underlying rule or knowledge base. No free sketches improvised by the designers are allowed as every instance of graphical expression is constrained or interpreted according the logical or functional attributes coded into the system. Given that architectural design activities are largely expansive in the sense that meanings of drawings or other design expressions may not be fixed even at the time when designers draw them, the highly restrictive approach of knowledge base systems has not gained continuous attention of a wider user community except some highly specialised design areas or products.

As a further development to those of the Expert Systems and Knowledge Base Systems, Case-Based Reasoning (CBR) systems were first put forward by computer scientists and AI researchers in the early 1990s (Kolodner 1993), and were subsequently applied by CAD researchers in various design fields. The cores of Expert Systems are the computational heuristics that fire particular rules in the systems and reach a set of solutions. CBR systems put more emphasis upon the construction of "cases" which are the representations of some past experiences or knowledge of solving particular problems. In terms of building a CBR system, the crucial thing is the "indexing" in cases that structures data in certain ways to enable case retrieval according to the end-user's specification of the problems at hand. The recalled cases can then be "adapted" into new solutions, which they themselves become further cases in the case-based systems. In building design, just to mention two CBR systems as examples, Maher and others at the University of Sydney developed a research system called CASECAD which combines a CAD drawing system with a case base for planning structural systems of various building heights (Maher et al. 1995, 141-164). Another research system called KICS was developed by Soon-Ae Yang, Dave Robertson and John Lee at the University of Edinburgh dealing with building regulations compliance according to the Scottish building codes (Yang et al. 1993).

3D Graphical Modelling and Rendering Systems. If one judges the success of an IT application in design practice in terms of the number of users, 3D modelling and rendering systems are certainly the most successful ones. CAD software developers (mostly software companies in the US) have produced modelling and rendering packages of industrial strength, targeting at end-users working in architecture, mechanical engineering, and civil engineering sectors. In the late 1990s, several CAD packages have achieved various degrees of worldwide popularity. Names like *AutoCAD, MicroStation, VectorWorks (MiniCAD), ArchiCAD, formZ,* and *3Dstudio* are no strangers to architectural students and practitioners worldwide.[2]

Multimedia and Architectural Design Presentation. Presentation of architectural designs to an audience (design critics at schools of architecture, or clients and users in the real world) is an important part of the design process. The use of computers for architectural design presentation and communication has now been widely accepted and expected (Coyne 1995). This may typically involve multimedia presentations that draw upon a collection of digital documents of various kinds: still or animated images, sounds, texts, audio/video clips and so on. Multimedia-based architectural presentation is made possible as, firstly, a range of software and hardware for generating and viewing documents of particular digital media has become widely available, and, secondly, special software for piecing together digital documents of different kinds into integrated packages has also been developed to facilitate "multimedia authoring." Given the extent of user-friendliness achieved, architectural designers themselves can produce multimedia documents effectively by using software like *MacroMind Director* or *Adobe Premiere,* in conjunction with other CAD modelling/rendering and digital imaging tools.[3]

Virtual Reality and Virtual Environments. The term "Virtual Environments" (VE) is closely related to "Virtual Reality" (VR) but often refers to different things to different people. Some people consider VE as computer-based visualisation environments employing components of VR technology, while others may mean the computer models of some existing or fictional built environments as the end products of using VR technology. One of the most distinctive developments of VR or VE is the innovative display and interactive technologies such as head-mounted data glove-based 3D immersive display or desktop panoramic viewing. Clearly, VR technology offers alternative modes of viewing and interaction to those of conventional single-screen based monitor technology, which may ultimately prove more pertinent to space-oriented applications like architectural design. However, we have not seen wide use of VR technologies being employed by designers directly as design tools. This is mainly due to the lack of two-way connections between the current VR systems and conventional CAD or CAAD tools, and most professional designers still rely on conventional modelling and rendering tools for sophisticated editing of any models they may have in mind. A likely immediate development is a connection between conventional CAD systems and VR technology to establish data communications between CAD modes and VR modes.

The review above regarding the research, development and application of computing systems in architectural design is only tended as a general backdrop for the main theme of this book on computer-supported collaborative design. There are other areas of design computing not mentioned above which does not imply that they are less significant. New ideas and areas of CAAD development are constantly emerging due to different understanding of design processes and newly available computation techniques and digital technologies. To get a view of the latest ideas of system building and applications, a useful resource to look into is the major international CAAD conferences taking place regularly around the globe, for example, CAAD Futures, eCAADe (Europe), ACADIA (North America), CAADRIA (Asia), and the SIGRADI (South America).[4]

1.2 CSCW: A social-technological perspective of practical computing

In a series of ACM-sponsored conferences in 1984, the term "Computer Supported Cooperative Work (CSCW)" was first coined by Irene Greif and Paul Cashman. The themes of the conferences were to examine how people work together and how information technologies may support cooperative working. The conferences led to the subsequent publication of the first book on CSCW, advocating CSCW as a distinctive research field (Greif 1988). In many aspects, CSCW presents an alternative paradigm of research and development of computing systems. Firstly, there were the attempts to define issues concerning how software and hardware systems can be made to facilitate multiple users who undertake some tasks collaboratively. Secondly, more social-scientific research methodologies were developed or introduced to the actual processes of system development: for instance, ethnographical study of group working, social linguistic models of human communication, user participation in system mock-up testing, capturing and reconstruction of human body languages, and so forth. As the field of CSCW developed, a social-technological perspective of computing and communications systems was opened up to address issues other than algorithmic performance or data structures design. Since its inception more than fifteen years ago, the CSCW research communities world wide have been active; major international CSCW conferences are held in North America (CSCW) and Europe (ECSCW) annually.

Design has been considered as an important subject area of CSCW as it is observable that design of large or complex artefacts is usually conducted by designers working as teams. One of the CSCW sub-areas has to do with the research and development of "shared drawing spaces." A fair amount of research has gone into how people work together through drawing activities. Being motivated and guided by observational studies of working group graphics and shared drawing space activities, researchers have implemented a range of prototypes of "collaborative drawing systems." To extend conventional single-user systems, experimental drawing tools are designed and implemented to be used by a group of users for undertaking collaborative design in various spatio-temporal situations (e.g., collocated vs. distributed and synchronous vs. asynchronous). "Collaborative writing" is another sub-area of CSCW that can also be related to collaborative design if design is considered in an even broader scope (the writing of constructional notes or specifications, for instance). In this book, however, we shall focus on the area of collaborative drawing or modelling in which graphic representation and expression play a major role in collaborative design activity.

1.3 Computer supported collaborative design: A new direction for CAAD?

As briefly mentioned above, the history of computer-aided design systems in architecture shows clearly that CAAD has not been developed in the context of group work. Given that computer networking is a relatively recent development, it is not surprising to see that design tasks have been typically seen and modelled from a single individual's point of view in most CAAD systems. In the late 90s, however, there appears a new set of agenda exploring the potentials of computing in supporting human communication and collaboration. Clearly, similar to many other applications, this emerging paradigm of CAAD is closely associated with the development in digital networking (hardware as well as software). In terms of network applications, we can now have the options of *Closed Net* or *Open Net*.

We can say that the early CSCW systems are examples of Closed Net applications. Dedicated networks for connecting a number of machines with multi-user interface were built to deliver shared working and information spaces. The net is closed in the sense that there is a specific (or fixed) boundary and architecture of the networking that cannot be extended without substantial system reconfiguration or scaling up. "Intranet," a more recent type of network application, can be considered as another type of closed net that aims to achieve a high degree of efficiency and security as required by corporations or safety-critical applications.

Given its current status, the Internet, on the other hand, is an Open Net that it can be accessed by anyone whose computers are connected to the Internet. This is not a proper place to reiterate the history of how the Internet was developed in the first instance and how it has evolved since its inception. In short, the Internet is comprised of software applications running among various hardware platforms connected through local area networking or "dial-up" networking on a global scale. The Internet has not been and will not be owned and controlled by a single governmental or corporate establishment; there is not a limit of how many servers and clients are allowed to be connected in the end. Currently, "traffic congestion" of the Internet due to the extraordinary growth in its uses is more of a problem to frequent net users.

The technology of Internet or Intranet is now highly relevant to CAD because a new generation of CAD systems has started to emerge. These new CAD environments will provide designers with system components that are "network-aware." Given these network-related functions, designers will work in a more dynamic CAD environment capable of facilitating communication and interaction with other sources of information or applications brought into by an intranet or the Internet. CAD system researchers and developers have begun to show how "uploading" or "downloading" information or application to and from a network can be related to architectural design processes.

1.4 Basic issues raised by architectural design

There are two main reasons why I consider that an investigation into teamwork in architectural design can contribute to an understanding of collaboration among people when designing things in general: Firstly, as a matter of technical necessity, the design of buildings, or, more broadly, of environmental artifacts, has to be based upon the participation of a group of designers with various expertise, each of whom is more capable than the others when dealing with a particular design domain; secondly, as a consequence of exchanging critical judgements among participants in the course of designing, the production of final unity in design products is a common concern shared by all parties of a design team.

To put the above problem context as perceived more clearly, let us consider a picture of design practice in focus. In the world of architectural design, we often see two salient features in the entire practice of design production, which may be better illustrated by the diagram shown in Figure 1.1. On one hand (the left-hand side), there is the participation of multiple individuals of various design perspectives and expertise, who may act on heterogeneous design worlds to undertake domain-specific design tasks. On the other hand, there is the goal of developing a single integrated design that satisfies the design requirements elicited by all design participants. This overall picture of collaborative architectural design processes seems to prompt some basic issues of developing computer support for design collaboration:

Design Integration. To support the representation and flow of design concepts, ideas, and information generated by members of a design group such that domain-specific partial designs can be combined and evolved into coherent integrated schemes that will meet design objectives requirements of various domains. The integrated designs as sets

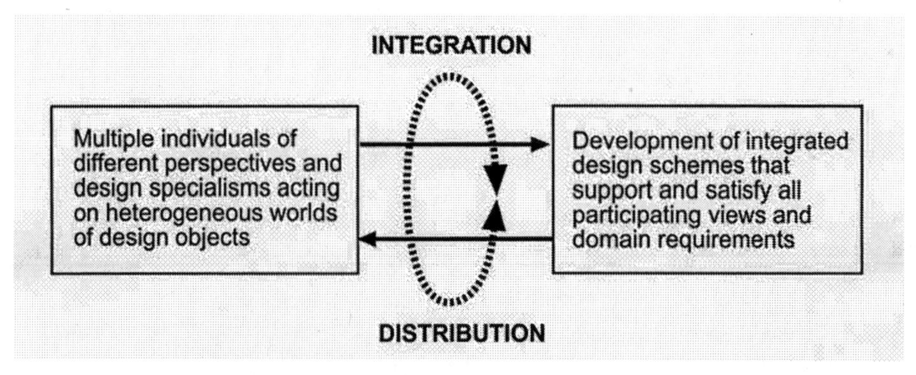

Figure 1.1 Basic issues raised by collaborative design in architecture.

of drawings or models are further used by the project team to communicate with other people involved in the project such as clients and builders.

Distributed Design. Collaborative development of integrated design as described above may lead to new initiatives and plans of how parts of design should be distributed among members of the design team because of the emerging wholes. As a design project develops, the part-whole relations are under constant interpretations by different team members from different design perspectives. Computing facilities can be developed and deployed to support dynamic distributed design by recording and maintaining new working relations as defined and redefined by team members during a project's lifetime.

Interplay between Integration and Distribution. Having pointed out integration and distribution as separate aspects of collaborative architectural design, there is a further level of computer support to be considered: the interaction between integration and distribution. In later chapters, I shall show that design integration may lead to new runs of design distributions, and vice versa. That is, there can be a reciprocal effect of how integration is interrelated with distribution arising from constant reinterpretation or negotiation of parts-and-whole relations. However, given the creative and open-ended nature of architectural design it is neither meaningful nor possible to propose any reliable models of how integration and distribution may react upon each other. In many ways, design is better characterised by spontaneous occurrences of putting parts into wholes as well as breaking existing wholes down into smaller parts whenever the design teams feel the need to do so; at one time, some designers see integration leading to distribution, while at another moment, it is more a matter of making sense of distributed parts by way of converging them. It can be argued that collaborative architectural design cannot be truly supported without addressing the interplay of integration and distribution as a totality of group design processes.

1.5 Design as modelling complex objects

In investigating computer-supported collaborative design, I take a rather different viewpoint of design: design is viewed as an activity of modelling complex objects. Design as a creative human activity may lose its richness if we try to characterise the features of design via a specific singular perspective, for instance, information gathering or processing, knowledge representation, actions of making, or performance engineering. Most people would agree that no single one of these could be said to be more representative or more important than the others. Design as modelling complex objects is intended as an alternative framework for describing design as it is explained in more detail below.

In the world of building design, owing to the large varieties of components as well as the dynamic relations between parts and wholes, information about the objects or artefacts as designed by architects and/or engineers tend to be complex in nature. In the long tradition of building design, there have been techniques devised to represent and manage the complexities involved as we can observe in an architectural or engineering workshop. Designers often devise and manipulate physical tokens (e.g., cardboard, strings, paper, wooden blocks and polystyrene, etc.) with which they construct physical models for various purposes. Working with these physical models, designers are better facilitated to develop, reflect, and communicate their stages of works with others. Though it can be extremely tedious, model making has been generally considered by design educators and practitioners as an essential part of design processes. The construction and use of physical models has been widely observed not only in individual cases (see, e.g., Goldschmidt 1988; Janke 1978) but also in group processes (see, e.g., Schön 1985; Ward 1987).

In addition to making physical models, designers also produce various forms of drawings. There are intimate relations between the production of drawings and model making. Sometimes models are made after drawings have been produced (i.e., drawings serve as blueprints in model construction); and at other times, drawings are produced on the basis of models constructed (i.e., models serve as referents or primary sources for drawing construction). In most cases, designers work in a mixed manner; that is, designers produce drawings to develop or elaborate design solutions as suggested by model construction, and designers construct models to better inform themselves of the consequences of associating or disassociating parts of a design as explored in drawings. It is therefore reasonable to assume that, in designing complex objects like buildings, there is constant information flow between drawing and model making, which may or may not be recorded explicitly. In this regard of generating and managing design information, drawing can be considered as a somewhat abstract form of model making.

As already pointed out sometime ago by Tjalve and his colleagues in their systematic treatment of "engineering graphic modelling" (Tjalve et al. 1979), a drawing is a model if it is made to demonstrate the following properties:

- A drawing represents *modelled properties* (e.g., structure, form, material, dimension, surface, etc.);
- A drawing has a *receiver* who is the person or persons to whom the drawing communicates information;
- A drawing is *coded* in systems of symbols (e.g., coordinates, graphical symbols, types of projection) known to the receiver.

Seen from design as modelling complex objects, the activity of designing in general can be described in terms of two interrelated aspects: the formation (or, formulation) of conceptual structures and the performance of modelling actions. The former decides to a large extent the range of basic constructs and their relationships in design use; the latter, when performed by individual designers, can lead to specific design descriptions (as in documents of construction specifications) and depictions (drawings or 3D models).

When connecting design with the uses of computers as media and tools, characterising design in terms of modelling complex objects may seem more appropriate for the purpose of pursuing academic research. In practice, we see that design conceptualisation does not normally take place during CAD operations. This phenomenon may be attributed to the fact that, historically, major commercial CAD packages have been developed mainly for design production. Using these packages, designers have no places in the systems to describe the concepts related to the design tasks at hand even if they wish to do so. If design conceptualisation is supported, we will have design tools in which the kinds of conceptual structures, as largely embedded in the processes of making physical models or drawings, can be made overt when designing with these tools. As an example, though developed purely as an academic research prototype, the MOLE (Modelling Objects in Logic Expressions) environment attempted to achieve just that (Krishnamurti 1985; Bijl 1987; Tweed and Bijl 1988; White 1992). In MOLE, a general descriptive formalism, "kinds-slots-fillers," was provided for designers to represent and evolve their own conceptual structures such that specific design drawings and specifications can be later instantiated. Although MOLE has not been further developed into a practical general purpose CAD system, the idea of achieving end-user constructible dynamic integration of design conceptualisation and design production remains a legitimate goal in the development of extensible CAD systems incorporating various approaches to end-user programming (see, for example, Nardie 1993; Aish 2000b).

Seeing design as an activity of modelling complex objects can also provide a fresh viewpoint to study design communication and coordination among multiple individuals. There are researchers who focus on the *activities* observable in collaborative design sessions (e.g., Bly 1988; Tang 1989). To meet the requirements of direct communication among collaborators, the provision of real-time computing supports for group interaction in design (i.e., via talking, gesturing, sketching, writing, etc.) among geographically distributed participants is of primary concern. However, to the CSCW community, collaborative design activity seems to centre on drawing. How people collaborate with each other through the acts of drawing has attracted considerable research efforts. Having read the CSCW research findings in collaborative drawing and writing, I consider that design as a modelling activity may still open up a different direction of enquiry that will elicit further requirements for computer-supported collaborative design, especially in the field of collaborative building design.

1.6 Observing and interpreting case histories

A collection of case studies of design projects is presented in this book. The design projects studied are *historical* in the sense that they were selected from published design documents and research literature that reports on *real* building projects undertaken by the designers working as teams. The materials gathered from the case studies are mainly designers' diagrams, drawings, illustrations, and models accompanied by the designers' or researchers' textual descriptions annotating the graphical materials. Case study or *protocol analysis* has been widely practiced in conducting research into design thinking, design processes, and design communication. (See, for instance, Rowe 1987; Bucciarelli 1988; Schön 1991 among others). In a more *laboratory-based* design study, a huge amount of data can be obtained by using tools like video cameras, tape recorders and so on to record and document what designers think and do. This may well capture more subtle information such as designers' hand gestures or even facial expressions. Designers as research subjects are invited to carry out some predefined design problems or tasks in rooms equipped with data-capturing devices. Drawings and other design expressions produced by the subjects as a part of important raw data are the outcome of designers' a few hours of work at most.

Case studies based on historical materials are of a different nature. In contrast with the kinds of contrived artificial controls imposed by most laboratory-based approaches, design representations and expressions presented in historical cases are the outcome of much longer processes that may last for months if not days. From a research point of view, since observing and interpreting what designers produce is an essential part of design studies, it is important to be aware of the contexts in which the research is carried out. In individual as well as in teamwork cases, designers may naturally react to different conditions such as time-limit (e.g., hours vs. months) and the given working situations (e.g., being in simulation or being real), and produce design outputs of different natures that may reveal what design is about very differently.

As said, the design studies presented in this book follow a historical reconstruction route based on the evidence of designers' outputs observable in drawings, models and, occasionally, their verbal recollections of what happened in the design processes. My aim of analysis is to derive an overall picture of what collaborative design in architecture is involved. The limitation of this approach is the lack of experiential or empirical details of what actually happened. The research conclusions reached are therefore not adequate enough to suggest any specific designs of system development. However, from the beginning of the research, I was quite clear myself that I did not consider myself as a CAD system developer of any kind. My attempt of observing and interpreting the cases of project histories was to arrive at general descriptions of the representation and communication requirements that may inform some kinds of system frameworks for supporting design collaboration. In many aspects, design representation is fundamental to design collaboration as it reflects a basic organisation of the sharing and distribution of design information in terms of parts and wholes. By analysing the relations among parts and wholes, we can at least make some plausible suggestions as how design communication is necessarily involved in design collaboration. It is for the research purposes that I make distinctions between representation and communication. In practice, designers' development of representation is intimately related to how they might want to communicate and coordinate with one another.

1.7 Approaches to research on CAAD

To develop digital mediums and computational tools, the richness and complexity of design activity and processes as appeared in many fields of design has been explored by academic and industrial researchers around the world. As a member of the CAAD research community, I believe that designers' digital palettes and modelling environments will certainly continue to evolve in the years to come if there are continuous inputs from research. In my observation of research on CAAD in general, there can be two basic types of approaches to doing research: *technology-pushed* and *practice-pulled*.

Technology-pushed research is interested mainly in introducing and experimenting with new computing components and infrastructures in some domains of design. A new generation of CAAD systems may emerge because of newly arrived computing platforms or networking infrastructures, if they are tried and accepted by the designers as the end-users. Taking the most recent examples, due to the availability of object-oriented technology, the Internet and the World Wide Web, major CAD developers such as *MicroStation* and *AutoCAD* have produced a new generation of CAD tools that can handle information sharing for instance. Being optimistic about it, new technologies can simply open up new applications of CAAD that have not been envisaged before. In this case, new applications are brought up not necessarily because of preoccupied problems or requirements of architectural design.

On the other hand, practice-pulled research starts with better understanding and describing design activities and processes that may or may not contribute directly to the development of new design computing environments. We can find many good examples in the well-known journal of Design Studies. Just by looking at design practice, research can be developed along the lines such as design products, processes, procedures, activities, theories and histories of design, or even evaluation of design computing tools. Like design itself, design studies is almost a "multi-lingual" discipline, presenting results in a wide range of perspectives: psychological, sociological, mathematical, linguistic, to name a few. Given that design practice around the globe is going digital at large, there has been a strong tendency of conducting design studies for the purposes of developing computational systems. In this respect, practice-pulled research is essentially a requirement elicitation and analysis activity.

It seems that we now have two major resources for research into computer-supported collaborative design. One is from *collaborative computing,* which can now be achieved through rapid combination of various computing and communication technologies, and the other is from *design computing* which already presents diverse approaches to the making of design tools and environments in CAD. As pointed out by Jonathan Grudin (Grudin 1991), collaborative computing utilises networking, communications, concurrent processing, and windowing environments. These new techniques and facilities have opened up the possibility of collaboration via group interaction in real-time even if participants are geographically remote to each other. To better understand the earlier issues and ideas of collaborative computing, especially in the designs and implementations of group drawing tools, a whole chapter is devoted to a survey of this technological development in the pre-Internet era.

Working closer to some domains of design practice, some CSCW researchers already considered that the sharing and use of certain kinds of structures of design artefacts by group members can be captured in a formal language or a dedicated design system (e.g., Fischer et al. 1992; Lakin 1990). Collaboration-supporting design tools developed in this view contain an underlying assumption that the structure and knowledge of the design products built into the systems is stable and usable to most design practitioners over the lifetime of the systems.

Research in theories of design computing has shown a different approach to that of knowledge-based ones. Again, referring to the MOLE experiments, it has been shown that designers can work with a general descriptive system in constructing their own models of design objects in pictures and words. As proposed by Bijl (1986; 1989, 172-208), a more fundamental (or, rather radical, as some of us may consider) position is to allow users to be able to instruct machines what to do in a CAD modelling environment with minimal or no prescribed domain-specific knowledge.

In drawing possible links between the existing developments in collaborative computing and in (non-prescriptive) design computing, a theme of study into computer-supported collaborative design is emerging: *exploratory collaborative design computing*. Given the background of enquiry, this book sets out to investigate the requirements for supporting design as an explorative process involving communication and coordination among a team of designers. It is expected that the requirements derived from the present study can serve as a useful set of pointers to future development of a collaborative design computing framework.

In my view, research into computer-supported collaborative design can be and should be approached from technology-pushed as well as practice-pulled. There can be potentially interesting connections between design studies and system research development though they may have not been undertaken by the same researchers. Research of technology-pushed kind will involve not only implementation of new tools but also usability testing of the new applications emerged. The importance of maintaining constant communication and interaction between technology-pushed and practice-pulled research is getting evident as time goes by.

At one time, the developers of CAAD systems were themselves experienced architectural designers. The knowledge about user requirements, system designs, and evaluations were all in the same heads and hands. Now, due to the continuous sophistication of computing technologies, it seems that people trained as architectural designers are in a better position when conducting design studies and eliciting system requirements, while people trained as software developers are better equipped to attain integrity in system design and development. Clearly, dialogues between the two professions are crucial in shaping future computer-augmented design practice. Following my own limited experiences in architectural design, I often aspire to the task of conveying my understanding of design activities and strategies to system developers and of explaining system behaviours to users of design computing tools. Perhaps in many places, this book reflects my tendency of doing so.

1.8 Scope and organisation of the book

In the course of my studies over the years, meeting with other colleagues and conference participants, I keep stepping on new lines of enquiry and approaches to system developments. Architectural design and the processes of realisation of building projects are enormously rich and complex which defies systematic abstraction and formalisation. Knowing that there can be many further topics of design collaboration included in a book like the current one, I consider it necessary to limit the scope of discussion to the collaboration among members of (professional) design teams. The obvious omissions are designers communicating with their clients/users, and constructors/suppliers, which in reality cannot be separated from communication within the design teams. To include all of these aspects of design collaboration in a single volume is certainly desirable if not too ambitious, but the complexity may render its readability low. The issues of design collaboration among participants *external* to design teams in a building project, such as clients, users, and builders, deserve dedicated volumes of research and writing. Attempting to achieve a degree of coherence and conceptual clarity, I limit the scope of the current volume to the aspects of collaboration *internal* to building design teams.

Contents of this book are organised into the following chapters. Chapter 2 presents a survey of the recent experiments in supporting collaborative drawing and design activities. We shall look at how the issues of supporting collaborative drawing and design have been identified and responded with system designs put forward by CSCW researchers worldwide. Firstly, there are interesting findings of group interaction in multi-party drawing and design activities. I then introduce a framework of classifying the system design issues that the CSCW researchers have addressed, and thirdly, by correlating the patterns of group uses with the system features implemented, a catalogue is presented to give an overview regarding the current development and uses of prototype collaborative drawing systems. The survey shows that there were at least three different strategies for developing collaborative drawing support tools: as media of real-time graphical conversations; as tools used by participants to manage design ideas; and as media of performing 'team-room' activities. The differences reflect the existence of diversified understanding and technological responses to what and how human collaboration in design may be supported.

Without further delving into the technical details of recent CSCW technologies, in Chapter 3, I shall move onto design studies of team working in architectural design. The studies are based on the documents of several case histories of real building design projects, including a fountain project developed for the Seattle City Centre, the revamping of an industrial research laboratory in Indiana, a new college chapel built in Cambridge, a new headquarter complex for a telecommunications company in Coventry, and a private house design in Japan. These cases show that successful teamwork in architectural design seems closely related to certain common design concepts or metaphors discovered at some point of group working. The concepts or metaphors discovered then serve the team members as shared reference frameworks for combining distributed designs into emerging design solutions. The sharing of the common metaphors among team members interacts continuously with individual interpretations of what parts are and how parts are interrelated with one another to form overall architectural solutions.

Following the case studies on discovering design metaphors, a *bottom-up* scenario of collaborative design is presented, which can be considered as an abstraction of the common features observed across the case histories. To be able to better describe the aspects of information flow in the bottom-up scenario of collaborative design, a *situation-theoretical* framework of description is introduced. The theory of situations, originated from the work by Jon Barwise and John Perry on formal semantics of natural language (Barwise and Perry 1983; Barwise 1989), suggests a framework for describing information flow and constraints on information flow, which I consider relevant to the discussions of collaborative design processes. Some of the basic ideas of the situation theory on the logic of information flow are briefly explained. On the basis of the situation-theoretical framework, I then develop a more detailed exposition of the bottom-up pattern of collaborative design.

As a sequel to Chapter 3, Chapter 4 continues to investigate the emergence of common images and metaphors. Following the situation-theoretical exposition given earlier, I describe further the constraints on the flow of information among the types of situations elicited from the bottom-up pattern of collaborative design. Situation types are conceptual constructs that I devised to describe an internal structure of design communication in terms of modelling acts, modelling spaces, and various design states identified from the case studies. It is by examining the constraints on the flow of information among the situation types derived from the workflow pattern that I conclude the constraints underlying the bottom-up approach to collaborative design. The study of the constraints is useful as they are the pointers to a set of issues regarding how collaborative design through discovery of common images and metaphors may be better facilitated by computing communications. In particular, I infer from the current study that online access to digital visual references can be useful to designers when communicating and explaining discovered common images or metaphors. The ever growing World Wide Web and other online visual libraries can potentially serve design teams as catalysts in communicating collaborative design discovery. Finally, I discuss the prospects of intranet and Web-based technologies in developing a computer-supported collaborative design environment encompassing some of the key features of information flow in design collaboration.

Starting with a further two case histories in building design, Chapter 5 presents another descriptive theory of collaborative design that differs from the bottom-up pattern described before. The first case is a study of the Colonia Güell Church design project in Barcelona. The evidence gathered from the project history shows that some kind of flexible generic framework was employed by a design team to sustain group dynamics through a project's lifetime. More instances of generic flexible structures are provided in the second case study, which is a review of the Taxis schemas, the spatial rules governing the subdivision of architectural spaces in the traditions of European classical architecture. Based on the case histories, a *top-down* scenario of collaborative design is presented in which I outline the important properties of flexible generic structures. A detailed exposition of the top-down pattern of collaborative design is developed in accordance with a situation-theoretical framework. The term "structures" is used extensively in my description of the top-down pattern of collaborative design, yet it did not bear any direct connection with the "structuralist" in Structuralism initially. In my later review of Herman Hertzberger's recent writing on architectural Structuralism (Hertzberger 1991), I realised that some interesting correspondences can be made between the structures in architectural Structuralism and the structures seen in the top-down approach to collaborative design. The correspondence found highlight the properties of flexible generic structures with which group dynamics in design collaboration can be associated.

Chapter 6 is the second half of studying the representation and communication in collaborative design involving the use of generic flexible structures. Following the situation-theoretical explanation in Chapter 5, the constraints on the information flow among the situation types identified from the top-down scenario are derived. Again, the constraints are the pointers to a set of issues with regard to how collaborative design in the top-down scenario can be better facilitated by computing communications. In discussing potential strategies of representing flexible generic structures in computational systems, the chapter includes an introduction to constraint-based graphics and other related research in computer graphics. By drawing on the group processes elicited from the top-down pattern, a system framework for developing a constraint-based object-oriented graphic modelling and mediating system is proposed, which aims to facilitate multidisciplinary collaborative synthesis of built form. To round off the current studies of collaborative design processes, I summarise some of the differences and connections between the bottom-up and top-down patterns of teamwork in architectural design. To take a more holistic stand of supporting teamwork, the distinctive patterns of workflow in collaborative design may be not necessarily implemented in two separate computer-based modelling and communication environments. We may speculate a pragmatic approach to define software components by focusing on a core collection of spatial operations and image editing tools, and the software modules can be configured and reconfigured by users to respond to different patterns of information flow.

Perhaps being pushed by the recent widespread and acceptance of the Internet and World Wide Web (Web, in short) technologies, developers of industrial CAD software have also started to offer features related to team working in building design. Chapter 7 presents a review of three popular commercial CAD platforms for collaborative working: ArchiCAD for TeamWork developed by Graphisoft, AutoCAD 2000i by Autodesk, and MicroStation ProjectBank by Bentley Systems. Unlike the development of research prototypes for collaborative drawing seen earlier, these new multi-user platforms have not been developed completely from scratch but extended by add-on components that are compatible with existing product lines. The commercial CAD software developers seemed mainly concerned with (a) how an assumed pattern of workflow can be implemented on the existing platform, and (b) how the proprietary CAD file formats can be still publishable and viewable over the Web. The products reviewed provide different strategies and solutions to these issues. Perhaps being more important than any specific platform developments, there is an emergence of industry initiatives on creating open-access electronic data schemas for pan-industry information sharing and exchanges over the Internet using the eXtensible Markup Language (XML). We shall look at examples of published XML schemas and discuss its implications for further industry development of collaborative design computing.

Finally, Chapter 8 is devoted to a discussion of further research on groupware technologies for collaborative design. Based on my experiences of conducting the case histories study, I feel it is important to establish a better research methodology by employing proper digital media and processing tools. The idea is to develop a *hypermedia case bank* for design studies. In pursuing practice-pulled design research, it is essential for researchers to be able to elicit raw data from field studies of collaborative design in action and store the data in as close to their natural forms as possible. The aim is to achieve a solid and versatile database on which research interpretations can be better grounded. The scope of collaborative design studied in the current volume is limited to collaboration among professional designers. In reality, collaboration in building projects is more complex, often involving clients/users and constructors/suppliers, to say the least.

To truly support collaborative design, we should look beyond system support for professional design teams to include clients and builders. This points to the setting up of *project-wide networking* and *online architectural services* such that professional designers, clients and builders can sustain effective project-oriented communications by supplementing their face-to-face meetings with direct or indirect virtual meetings facilitated through project-wide digital networks. The CSCW tools surveyed in Chapter 2 were examples of early attempts of multi-user interface technologies. Since then, newer interface technologies have emerged to support more novel approaches to digital interaction ranging from responsive workbenches to cooperative workspaces. The prospects of innovative interface technologies for collaborative design are high in the years to come as we are fast approaching the age of post-PC ubiquitous computing. Interfaces based on 2D conventional computer screens will evolve into more versatile and responsive options implemented in 3D interactive media. While innovative interfaces are being developed to improve the qualities of sharing visual information in design collaboration, tools for knowledge management also need upgrading to be operated in line with the Internet—the backbone of project-wide networking. The proposition proposed for further development of groupware is that *software agents* may become parts of a project team. Looking at the emerging agent technologies, I shall discuss several aspects that software agents can be authored to take some roles of managing information and knowledge in collaborative design. Finally, we need to look at design computing education afresh in relation to groupware technologies. As new sets of concepts and skills of digital interaction and communication are required to enable networked design learning, design educators also need to explore innovative applications of groupware technologies to enhance and enrich traditional design teaching practices.

Summary

The chapter starts with a historical review of the developments of Computer-Aided Architectural Design (CAAD), Computer Supported Cooperative Work (CSCW), up to the most recent trend of Computer Supported Collaborative Design (CSCD). The review sets up a background for later more detailed discussions on groupware and team working in architectural design. The aims, objectives and methodology of the book are explained, including the scope and organisation of the book. The introduction shows that information technology (hardware as well as software developments) as applied in architectural design has evolved into the paradigm of networking that supports group processes of design. Early developments of CAAD technology focused on the issues of translating design tasks into computer programs that generated numerical and graphic data. The main objective of CSCW, however, is concerned with supporting users working cooperatively by harnessing the technology of computing communications. Seeing the design of complex artefact like buildings as essentially group processes, the researchers and developers of design computing were opening up new system developments by combining conventional CAAD and recent CSCW capabilities into the newly emerging application field of CSCD. Design in this book is to be understood as an activity of modelling complex objects consisting of many parts and wholes contributed by various members of a design team. To better describe the communications involved in design collaboration, it is necessary to examine the properties and organisations of how parts, wholes and their interconnections are represented within the contexts of project development.

2 Early Experiments in Supporting Collaborative Drawing and Design

In the field of Computer Supported Cooperative Work, the research and development of collaborative drawing tools on its own has been established as a distinctive subject area. Researchers have experimented with communication and computer tools for supporting various conditions of group working in graphical media. This chapter presents a survey of the early experiments in collaborative drawing support tools developed during the period of 1987–1993, that is, roughly prior to the explosive growth of the Internet and the World Wide Web. The main objective of this survey is to review how the issues of supporting design collaboration were first elicited and addressed in the developments of early research prototype systems. The survey is presented in three parts: (1) findings from the observations of group interaction in drawing and design activities; (2) a framework for classifying the system design issues as investigated by the prototype developers; and (3) a differentiation of the prototype systems by examining how the system features developed are related to group uses of these systems. The survey shows that there were at least three different strategies of developing collaborative drawing support tools, reflecting diversified understanding and varied technological responses to what and how collaborative working in design may be supported.

2.1 Introduction

Design as a human activity is pervasive; and designers often work jointly to develop usable and meaningful artefacts. The problem of how to develop communication and computer systems that can support collaborative design or problem solving has become an active research area, attracting researchers working on various perspectives. In his bibliographical survey of the research literature appeared in the field of Computer Supported Cooperative Work (CSCW), Saul Greenberg introduced the key word "shared workspace" (Greenberg et al. 1991; Greenberg et al. 1995). It is noticeable that a large portion of research work on shared workspaces has to do with building prototypes of collaborative drawing support tools. Being perhaps motivated and guided by earlier observational studies of *working group graphics* and *shared drawing space activities*, (see Bly 1988; Lakin 1983; Tang and Leifer 1988 among others), researchers and system developers have attempted to design and implement prototypes of shared drawing systems. These tools were intended to be used by a group of users to carry out direct or indirect graphical communication. The requirements for these systems are different from traditional single-user drawing tools that have been as popular as word processors. One of the major objectives of implementing shared drawing tools is to facilitate communication and coordination among participants in the course of creating and using technical (formal) or non-technical (informal) drawings.

However, different research groups around the world have observed, described, and analysed group interaction in design differently due to their research interests and perspectives adopted. There accordingly appears a diversity of understanding as well as assumptions as to what actually constitutes a shared workspace that will enable human communication and coordination in carrying out design tasks. Differences in the basic investigation into "How are drawings (as shared artefacts) and group drawing (as shared drawing space activity) related to collaborative design processes?" have resulted in varied technical approaches to "What is to be facilitated by shared drawing support tools?" Though a number of group drawing support tools have been previously reviewed (see, for instance, [Lu 1992]), this survey, by reviewing a wider range of group drawing tools, was intended to be more comprehensive. Instead of giving descriptions specific to particular system implementations, the objectives of this survey are (1) to identify the important aspects of understanding collaborative drawing and design activities with regard to the original studies that have been made, (2) to present a framework for classifying the design issues being investigated by the prototype system developers, and (3) to catalogue the properties and components of the prototype systems reviewed in the survey.

The remainder of the chapter is organised as follows. A discussion of understanding group drawing activities from various perspectives is given in the next section. Following the basic concepts of group drawing discussed, Section 2.3 presents a scheme of classification of the design issues emerging from the past experimentation in collaborative drawing systems. In Section 2.4, to give a clearer overview of the status of prototype systems development in group drawing, a categorisation that combines the various modes of system use with the range of system components is constructed. Finally, some topics suggested by the survey for further studies into collaborative drawing systems are discussed in Section 2.5.

2.2 Aspects of studying group drawing and design activities

Most developments of collaborative drawing tools were motivated and guided by the researchers' understandings of group drawing and design activities. It is therefore an appropriate starting point to look at what has been observed and characterised about these activities. Table 2.1 is a summary of nine such studies.[5] The studies of the nine groups were chosen because they presented original research perspectives and responded to the problems with various prototype solutions. The differences arisen here are significant in showing that it is quite possible to have very different angles when characterising what are involved in group drawing activities.

To better understand the various issues raised by these original studies and experiments, a more detailed comparative study is presented below, focusing on the aspects of events, information, tools, and ownership. These four aspects are relevant to the survey because they are the common set of conceptual issues addressed by the research groups to various extents. Other issues concerning more of the aspects of system design and implementation will be discussed in the next section.

Events: collocated vs. remote and synchronous vs. asynchronous

Since any collaborative drawing or design activity must take place in space and time, the

Research Groups	Fields of Observations	Group Size	Research Methodology	Important Findings	Research Prototypes
Lakin *et al.* (Lakin 1983; Lakin 1988; Lakin 1990)	Informal and formal working group graphics in engineering, geological survey, and computing	One or more facilitators with non-specified group size	A linguistic analysis of working groups' text-graphics images and manipulation	Spatial and temporal structures in manipulation of text-graphics	*vmacs*, and *Visual Languages for Cooperation*
Stefik *et al.* (Stefik *et al.* 1987a; Tatar *et al.* 1991)	Small teams of computer scientists engaged in face-to-face meetings	2—6 persons	The setting up of an experimental meeting room by networking PCs and a large screen	The effects of turn taking systems on collaborative work taken place in a meeting room	*CoLab* *Boardnoter* *Cognoter* *Argnoter*
Bly *et al.* (Bly 1988; Minneman & Bly 1991)	Informal drawing in computer user-interface design problems	2—3 designers	Video-audio protocol and a drawing events/ actions/clusters analysis	The uses of drawing surfaces in different collaborative settings	*Commune*
Tang *et al.* (Tang & Leifer 1988; Tang 1991)	Informal drawing in computer user-interface design problems	3—4 designers	Video-audio protocol and an action-function framework for protocol analysis	The Importance of the process of creating and using drawings	*VideoDraw* *VideoWhite-Board*
Ishii *et al.* (Ishii 1990; Ishii & Arita 1991; Ishii *et al.* 1992)	Informal drawing and computer generated images in computer systems design	2—3 designers	Intuitive understanding and building various versions of experimental prototype systems	The importance of integrating social protocols with shared workspaces	*TeamWork-Station* *ClearFace* *ClearBoard-1* *ClearBoard-2*
Fischer *et al.* (Fischer *et al.* 1991; Fischer *et al.* 1992; Reeves *et al.* 1992)	Formal coding systems and the structures of expert knowledge in the designs of kitchens and computer networks	not specified	Observing design experts in work, then developing a conceptual framework and a demonstration system	The integration of the design of the artefact (made in combined graphics and texts) and the communication among designers	*JANUS* *NETWORK-HYDRA* *XNETWORK*
Lu *et al.* (Lu & Mantei 1991; Lu 1992)	Multi-layer informal drawing in architectural layout design	2—4 designers	Generating requirements by mapping design behaviours onto skeleton design	Fifteen user requirements for a drawing system to support idea management	*CaveDraw*
Wolf *et al.* (Wolf & Rhyne 1991; Wolf & Rhyne 1992)	Informal small group meetings for the conceptual design of a "Clutter Collector Robot"	2—10 participants	Developing a pen-based prototype system, putting it in group uses, and getting detailed user feedback	The potential meeting process gains/losses, and the areas for aiding search and retrieval of information	*We-Met*
Brinck *et al.* (Brinck & Gomez 1992; Brinck 1993)	The remnants of office whiteboard conversations made for technical versus administrative uses	2—3 participants	Recording and analysing sketches, writing, and retrospective descriptions of conversations	A classification of whiteboard objects and their semantic properties in conversational use	*Conversation-Board*

Table 2.1 A table of nine studies of group drawing and design activities that have influenced their own or other prototype system developments.

patterns of events can be generally differentiated in terms of four basic spatio-temporal structures[6] (Figure 2.1).

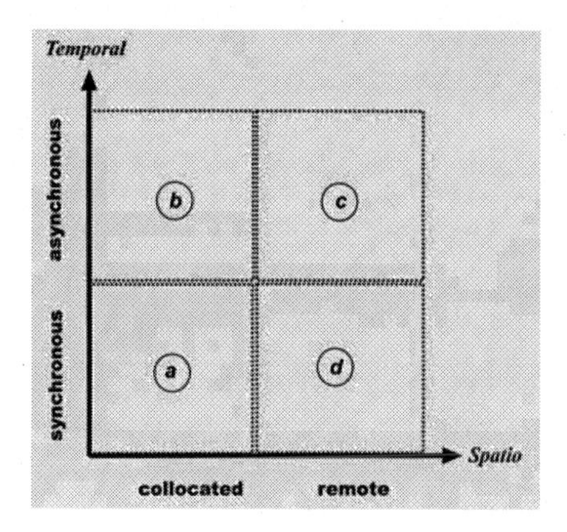

Figure 2.1 A spatio-temporal frame for classifying the events of group drawing or design activity into four basic patterns of collaboration. (The alphabetical order corresponds to the enumerated items described in the main text.)

a. *collocated synchronous.* All participants take part in a joint design session at the same geographical location via face-to-face simultaneous interactions. A typical group process of this pattern is a design meeting or a brainstorming session, which requires close physical proximity for intensive communication among group members. As studied by Fred Lakin (1983; 1988), working group graphics, operated by one or more operator (or facilitator), can play an important role in aiding collocated simultaneous interactions.

b. *collocated asynchronous.* Embarking the same project, members of a design group are located at the same workplace (say, a large design studio) through indirect communications over a period of time. Though not being a greatly detailed observational study, a scenario of what might be involved in a day of office work where five staff members took part in a telescope engineering project was described by Lakin (1990). What Lakin considered was that participants work in the same setting but are left alone to concentrate on different aspects of the project. It seems that "collocation" is not an obvious factor for "collocated asynchronous" being fundamentally different from "remote asynchronous" if the actual physical distance is taken into account. But, basically, "collocated" implies that members can see and talk to each other without moving from one place to another. To better illustrate this event type, we may think of a group of people engaged in "team room" activities. Another good example of collocated asynchronous is "shift work" (i.e., when one shift passes information to the next shift). In this manner, teamwork is carried out mostly via indirect communication among collocated participants (e.g., passing working documents or CAD files to one another not necessarily involving face-to-face meetings).

c. *remote asynchronous.* In carrying out shared design projects, participants work in geographically distributed settings through indirect communications. As a rationale of designing computer tools supporting design teamwork of this pattern, Gerhard Fischer and others explained that remote asynchronous collaboration could be commonly seen in modern technologically-oriented design projects (Fischer et al. 1992):

> Meetings and other types of direct communication are the commonly used means for coordination and collaboration, but in many situations — especially ones involving long-

term collaboration — these are not feasible. Modern design projects can extend over many years and can involve a high turnover in personnel. People who are not in the project group at the same time need to coordinate and collaborate in the design of a system.

d. *remote synchronous.* Geographically, group members are separated from a few feet to, perhaps, thousands of miles apart; but they are enabled to have direct communication while making drawings and notes. Events of this type attracted a majority of CSCW researchers' attention. In fact, a large proportion of the research prototypes was dedicated to supporting group drawing activities of this type. Under the spatio-temporal circumstances, collaborative drawing activities are made possible by providing participants with shared virtual drawing spaces which emulate as much as what can be achieved in a direct face-to-face interaction. In their separate experiments with prototype systems for remote synchronous group interaction in drawing and designing, some of the researchers have reported the following findings:
   - Drawing activities once shared (simultaneously) can pull designers together and increase their attention and involvement in the design task (Bly 1988);
   - The processes of creating and using drawings can convey information that is as important as, or, actually, not found in, the resulting drawings (Bly 1988; Tang and Minneman 1991a);
   - Designers are themselves skilled at coordinating communication, and the social protocols acquired via the face-to-face communication can serve the needs of constructive collaboration in the shared workspace (Ishii and Arita 1991; Ishii and Kobayashi 1992; Ishii and Ohkubo 1990; Tang 1989).

Information: action-oriented vs. representation-oriented

Collaborative work in design or problem solving has to do with the generation and exchange of information in one way or another. However, different perspectives have varied considerations of what information is to be captured and transmitted. On most occasions of synchronous collaboration, information useful to group interaction is considered to be *action-oriented*; that is, actions recorded or tracked by devices such as video cameras, shadow projection, mouse movements and so on, are the kind of information that participants may generate, recall, interpret, and share.

Actions that have been studied in group drawing activities include sketching, writing (listing alpha-numeric text), talking, gesturing, gazing (making eye-contact). The sharing of action-oriented information was believed to foster and maintain group awareness; but how is collective awareness related to the performance of collaborative work? Based on their work on the *ClearBoard* system (more details given below), Ishii and Kobayashi commented on how "gaze awareness," as supported by *ClearBoard*, affects two participants solving the "missionaries and cannibals" puzzle (Ishii and Kobayashi 1992):

> Through this experiment we confirmed that it is easy for the players to say which side of the river the partner is gazing at and this information was quite useful in advising each other.

On the other hand, synchronous or asynchronous sharing of design ideas or knowledge as conveyed by participants in some representational media is considered more important in less direct collaboration. When group work involves more technical matters (e.g., the production of technical drawings or the performing of graphical modelling in engineering design), collaboration may necessarily involve some formal system for constructing and interpreting individual or collective expressions. The views in favour of representation-oriented collaboration have presented the following points, arguing how formal representations of information (graphical as well as non-graphical) may serve group interaction in design.

In Lakin's view (Lakin 1986; 1990), spatial and temporal structures (schemata) can be observed when designers create and manipulate text-graphic expressions. These structures, on one hand, limit the kind of expressions and arrangements one may make, while, on the other hand, provide the basis for well-defined spatio-temporal regularities available for automated machine interpretation. Various *visual languages* for collaborative working such as co-authoring, brainstorming, and task structuring can therefore be formally defined.

In the view of Gerhard Fischer and others (Fischer et al. 1992), long-term indirect collaborative design takes the model that coordination of individual work in groups is achieved by the individuals' interactions with "group memory," which can be represented formally as

> ... a collection of shared information repositories containing a cumulative record of rationale, solution components, information about prior projects, and other information resources for collaboration.

Given the group memory represented both in (hyper-) textual and graphical forms, indirect communication among designers can be supported by the computer-based methods of "argumentation"[7] and "critiquing"[8].

Tools: homogeneous vs. heterogeneous

The third aspect of understanding collaborative drawing activities is concerned with the use of tools. The problem lies in whether all participants work with the same set of tools, or each of them may need to operate different sets of tools. There appears again a dichotomy between homogeneous and heterogeneous sets of tools used by participants. Heterogeneity of tools may be due to, for instance, being manual or computer-based, the structures of drawings produced, ways of storing and retrieving data, actions involved in manipulating expressions, or domains of interpretation and so on. Referring to Tang's, Bly's, and Fischer's studies (see Table 2.1), I can draw up two main arguments why the provision and use of homogeneous sets of tools are considered as being sufficient for supporting collaborative drawing and design:

1. Synchronous group interaction does not involve domain-specific information or knowledge; i.e., participants converse with each other on design issues readily supported by sufficient common sense such that "pencil and paper" type tools can satisfy the communication needs of the group.

2. Participants come from more or less the same professional background and work within the same design domain; i.e., there is no need for streams of expert knowledge across different design disciplines in the course of collaboration. It is therefore sufficient for all participants to operate the same set of tools, even if it is a highly sophisticated one.

Taking a rather different view, Ishii and Miyake introduced the concept of "open shared workspace," expressing that "group members should be able to use a variety of heterogeneous sets of tools (computer-based against manual tools) in the shared workspace simultaneously" (Ishii and Miyake 1991). Again, according to Lakin's observation (Lakin 1990), in ordinary office work, there exist *group-individual mode switching* (i.e., participants sometimes work as individuals and sometimes as members of a group) and *technical task switching* (i.e., work changes between various tasks requiring special technical support). The design of various visual languages for cooperation was aiming to provide heterogeneous analytical tools, to which not all participants need to pay equal attention. Lakin also believed that the availability of a general-purpose text-graphic editor, together with multiple special-purpose analysis tools would enable the team to switch between discussing general issues and dealing with more technical details during a meeting session.

Ownership: group vs. individual

There can be no groups without individuals, and this is largely true even if group members have the same background and work with common languages and tools. As a need or an obligation, an individual's identity is basic to the concept of ownership in any context of group work. A drawing created by a group member may not necessarily be owned by the individual, if other members are allowed to change or remove it at will. Shown by the observational studies, some researchers suggested that participants themselves are good at coordinating individual activities so that there is no need to provide extra facilities for controlling ownership; some others implicitly or explicitly addressed the issue of supporting the preservation of ownership control to prevent potential malicious or accidental acts such as removing individual or group work results.

Borrowing from the model of "permissions" found in many multi-user operating systems, the levels or degrees of ownership can be defined in terms of two dimensions: identities and operations. For identities, these can be further divided into, for example, personal, sub-group, group, all; in operations, there can be a differentiation between read, write, and execute. This file-based permission model, however, can only partially illustrate the ownership issue in a shared drawing space; a more fine-grained framework is needed. As reported in Lu's study of teamwork in architectural design (Lu 1992), three user requirements were identified, revealing the need for "seamless and dynamic" transitions between group and individual ownership:

> Allow participants to declare any portion of a sketch as private and not subject to deletion by others.
>
> Allow participants to identify, with no additional interaction, who owns a specific design sketch.

Allow participants to bring in their own ideas from a private drawing surface provided by the shared tools or from a private file to a shared drawing surface.

Given the different considerations of how ownership may be defined and managed, there were three kinds of approaches attempted: (1) the building of multi-user interface components, (2) the design of arbitration algorithms, and (3) ownership embedded in separate drawing spaces. To give some examples, the *CaveDraw* drawing surface, developed by Lu and others (Lu 1992; Lu and Mantei 1991), distinguishes between: "pencil" input (for producing pencil marks that can be changed by any participant); "marker" input (for producing marker marks in distinctive colours owned by individuals). The operation of "cut-and-paste" was further provided to enable transitions of design ownership (from private to public) during collaborative working sessions. For instance, a person can cut parts of a drawing that were originally drawn in his or her own colour marker, and then the parts are pasted into the public domain shown in pencil marks (Lu and Mantei 1991).

In arbitrating the potential conflict of multiple users grabbing the same drawing object simultaneously, the design of *GroupDraw* defined ownership in different levels ("all of us can see and touch," "I can touch but you cannot," "only I can see it") (Greenberg et al. 1995). Due to the technologies used in video-based or fused computer-video shared drawing spaces (e.g., *VideoDraw*, *TeamWorkStation*), since no one can change or erase other participants' work simply by viewing or pointing at them on one's own drawing surface, a participant naturally owns what he or she draws on an individual screen or desktop surface. With this approach, ownership is, so to speak, inherent. To summarise the above discussions, Table 2.2 gives an overview of these aspects of understanding shared drawing space activities. It shows that some research aspects are comparatively less explored either empirically or conceptually than the others.

2.3 Prototype developments and system features

I have reviewed the early research efforts on aspects of group drawing or design activities. Motivated or guided by various understandings of what group drawing and collaborative design were about, CSCW researchers have worked on the problem of how to support these activities through the development of prototype systems and the demonstration of using these tools in various contexts of group working. In this section, a survey of the system issues arising in these prototype developments is presented. As a result of this survey, the system design issues can be classified into five clusters: (1) structures of graphics, (2) network configurations, (3) information storage and retrieval, (4) multi-user interfaces, and (5) other dialogue channels. I should point out that the present classification, as derived from my comparative study of the prototype systems developed, is not intended to be exhaustive. Nevertheless, in my view, it is in addressing these issues that components of experimental collaborative drawing systems were introduced and put together by the system designers. A discussion of these issues is given in the subsections below.

Research Aspects \ Research Groups		Lakin et al.	Stefik et al.	Tang et al.	Bly et al.	Ishii et al.	Ficsher et al.	Lu et al.	Wolf et. al.	Brinck et al.
Patterns of Events	collocated & synchronous	Addressed to an Extent	Addressed	Not Addressed	Addressed to an Extent	Addressed	Not Addressed	Not Addressed	Addressed	Addressed to an Extent
	collocated & asynchronous	Addressed to an Extent	Not Addressed	Not Addressed	Not Addressed	Not Addressed	Addressed to an Extent	Not Addressed	Not Addressed	Not Addressed
	remote & synchronous	Not Addressed	Not Addressed	Addressed	Addressed	Addressed	Not Addressed	Addressed	Not Addressed	Addressed
	remote & asynchronous	Addressed	Not Addressed	Not Addressed	Not Addressed	Not Addressed	Addressed	Not Addressed	Not Addressed	Not Addressed
Informa-tion	action-oriented	Addressed to an Extent	Not Addressed	Addressed	Addressed	Addressed to an Extent	Not Addressed	Addressed	Addressed	Addressed to an Extent
	representation-oriented	Addressed to an Extent	Addressed	Not Addressed	Not Addressed	Addressed to an Extent	Addressed	Not Addressed	Not Addressed	Addressed
Tools	homogeneous	Not Addressed	Not Addressed	Addressed	Addressed	Not Addressed	Addressed	Addressed	Addressed	Addressed
	heterogeneous	Addressed	Addressed	Not Addressed	Not Addressed	Addressed	Not Addressed	Not Addressed	Not Addressed	Not Addressed
Owner-ship	group	Addressed to an Extent	Addressed	Not Addressed	Addressed	Not Addressed	Addressed	Addressed	Addressed	Addressed
	individual	Addressed	Addressed to an Extent	Addressed	Not Addressed	Addressed	Not Addressed	Addressed	Not Addressed	Not Addressed

LEGEND: Addressed; Not Addressed; Addressed to an Extent

Table 2.2 An overview of some aspects of understanding shared drawing space activity investigated by the different research groups in the survey.

Graphics primitives and operations

Drawings as visual objects, created and passed around among people, are often constructed from some graphics primitives, which may or may not have computational representations within a drawing system. In computer-based drawing surfaces, users are provided with drawing functions for making marks or constructing graphics objects such as lines, rectangles, circles, etc. The provision of drawing primitives and functions determines the properties of a drawing surface to a great extent, since these primitives delimit what pictorial expressions are allowed, and what drawing operations can be applied to parts of these pictures. Certainly, the design of shared drawing spaces is not an exception to this basic principle; but the requirements for supporting possibly concurrent multi-party interaction in a shared drawing space have motivated different views on what drawing primitives and operations should be provided. Five different structures of group graphics were found on this basic issue.

1 *Video-captured images of freehand sketches*. By using markers directly on drawing

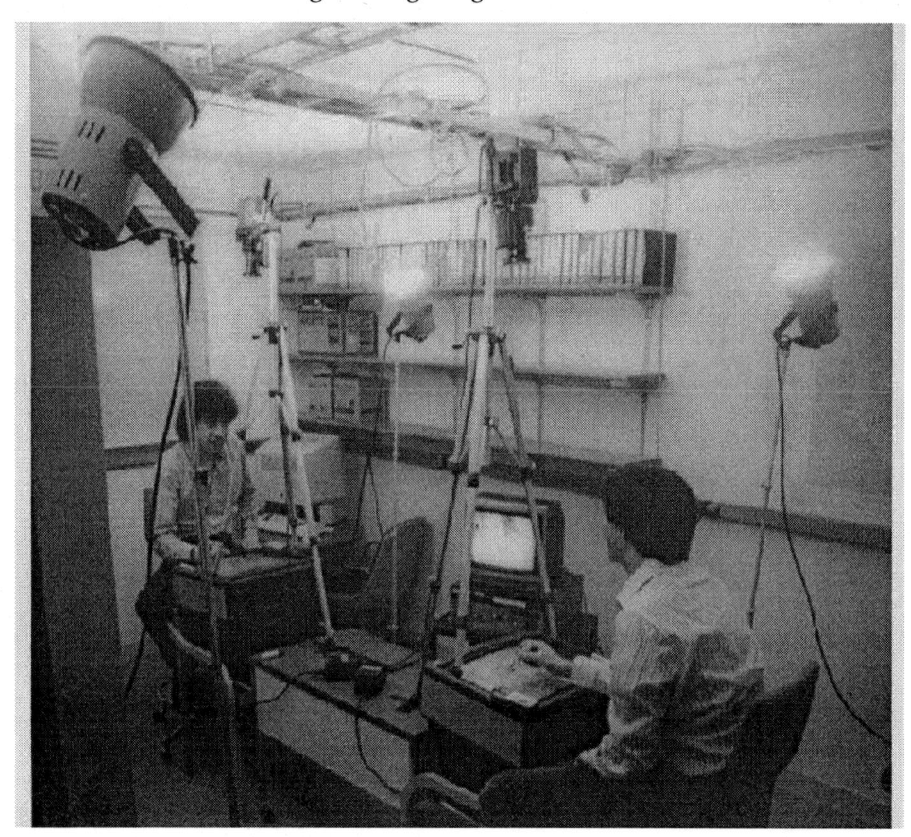

Figure 2.2 The drawing surface of VideoDraw, developed at Xerox PARC, used horizontal 20" video monitor screens with dry-erase ink markers (from Tang and Minneman 1991a, with the permission of the authors).[9]

surfaces, drawings are simply participants' freehand sketches. To transmit the images of sketches made by team members located in different workspaces, video monitors, video cameras, projectors, and networks are set up as working suites. Since there are no computational representations of drawings involved, participants can use white board markers to draw freely whatever they want. Supported by video networks, what appears on an individual's drawing surface is a synthesised visual space containing a translucent overlay of his or her sketches and video-captured images of those by others. Apart from physically erasing and taking pictures of the marks left on the drawing surfaces, little can be done on the drawings once made. Figures 2.2 to 2.4 show three design examples of (purely) video-networked drawing surfaces.

2 *Pixel-based graphics*. Pixel-based (or bit-mapped) graphics is often defined as the picture representation of drawings as arrays of pixels on computer screens, which

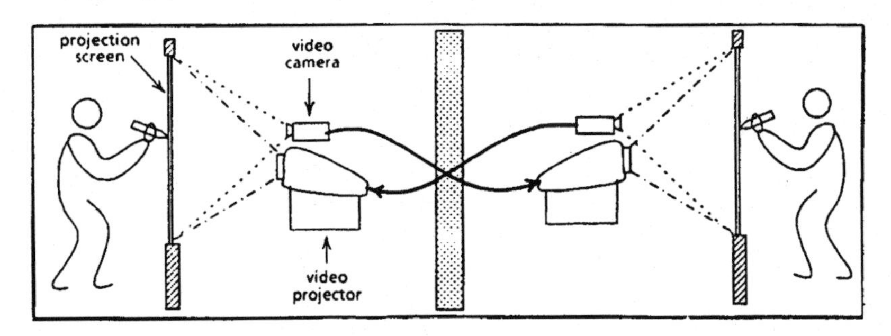

Figure 2.3 The shared drawing space of VideoWhiteBoard, also developed at Xerox PARC, used wall-mounted rear projection screens (approximately 4.5′ by 6′ with standard dry-erase whiteboard markers, from Tang and Minneman 1991b).

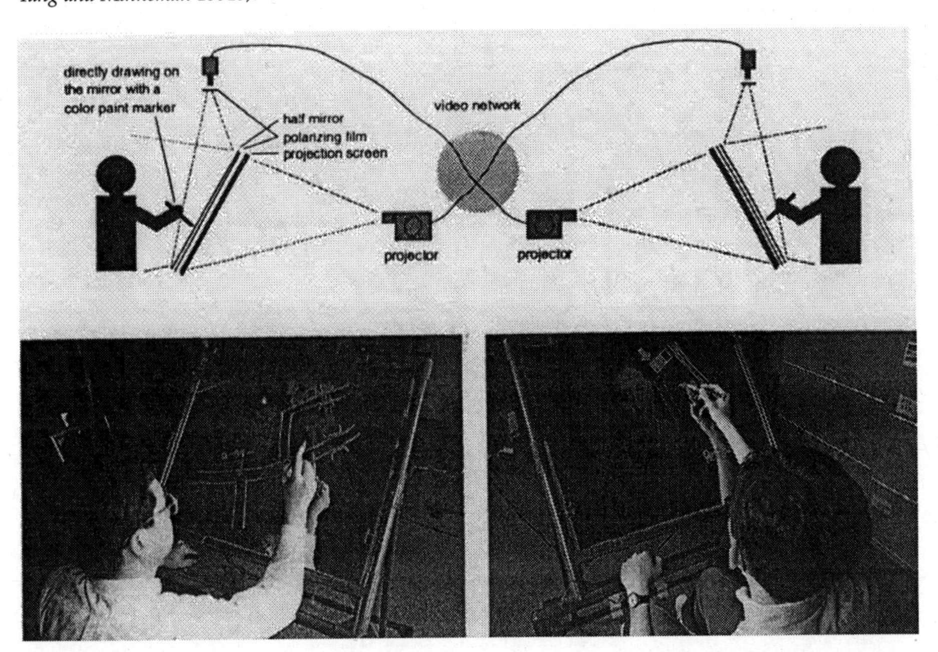

Figure 2.4 The shared drawing board of ClearBoard-1 developed at NTT was composed of a projection screen, a polarising film and a half-silvered mirror with water-based fluorescent paint markers7 (from Ishii and Kobayashi 1992).

corresponds closely to the data storage pattern in computer memory. As the primitive of most painting systems, a pixel has only the status 'on' or 'off,' and it has no relation to the status of others. Therefore, drawings in pixel-based graphics have typically no

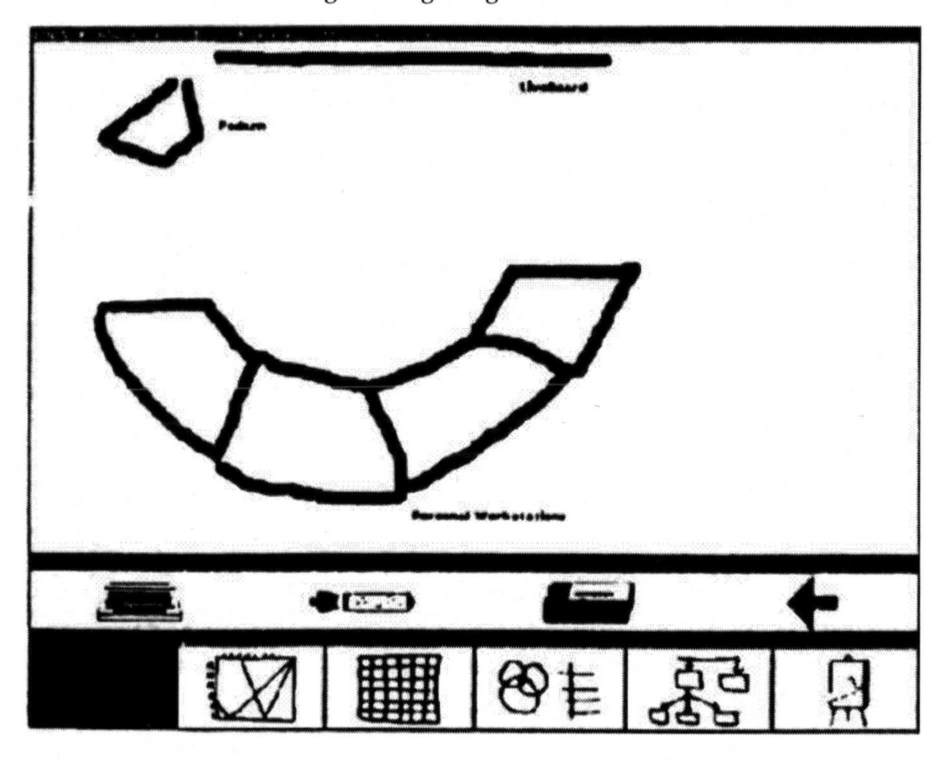

Figure 2.5 *Boardnoter of Colab at Xerox PARC provides participants with mouse-driven cursors ("chalk"), operating at individual workstations but not visible on the large meeting room screen (from Stefik et al. 1987).*

underlying structures or models specified by the graphics system. Drawing functions can be implemented as procedures for generating images of arbitrary marks, lines, rectangles, circles etc. Due to its simplicity, pixel-based graphics has been used in several prototypes of shared drawing space. Operations like erasing, selecting-and-dragging, and cutting-and-pasting are commonly used for changing bits of drawn or textual expressions. Among the shared drawing tools built of pixel-based graphics, a diversity in the design of input devices shows different approaches to experimenting with how drawing events can be shared among participants (Figures. 2.5 to 2.8).

3 **Object-structured graphics.** Freehand sketches and pixel-based graphics are unstructured graphical expressions to which very few operations can be applied. To be able to manipulate parts of a drawing as the constructs of line, rectangle, circle, etc., geometric structures need to be included in the implementation of graphics primitives. The term "object-structured" as applied to graphics refers to a system's drawing primitives being programmed as objects and stored in a database to be

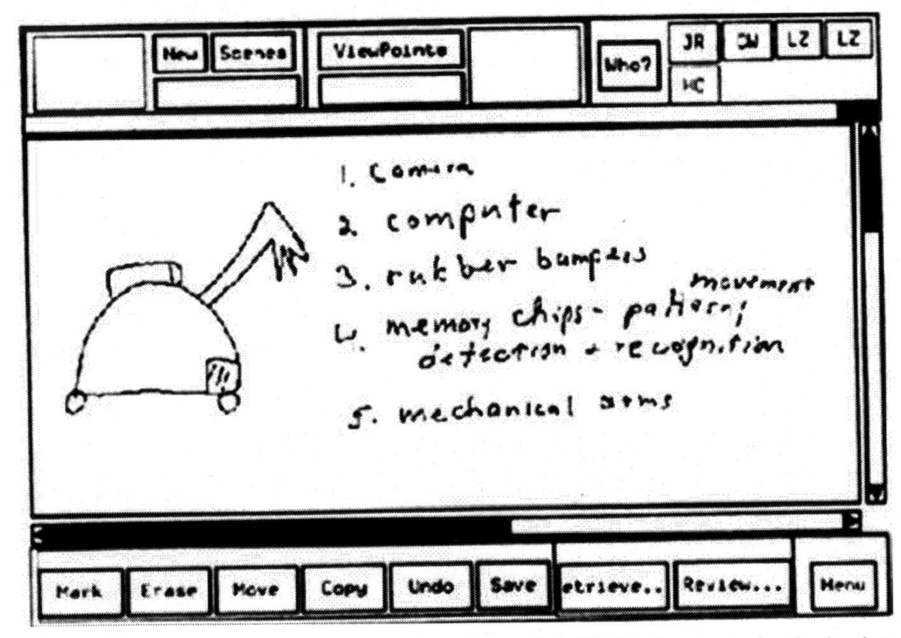

Figure 2.6 *Being similar to Boardnoter as a meeting support tool, the We-Met drawing surface developed at IBM Watson Research Center has the feature of "pen-based" interface (from Wolf et al. 1992).*

addressed, manipulated, copied as individual entities. In an object-structured graphics system, types of drawing objects can modify themselves with various sorts of operations, such as creating, moving, resizing, grouping, rotating, duplicating, deleting, etc. To give an example, one of the "tool palettes" of *Conversation Board* (Brinck and Gomez 1992) provided a range of geometric objects including oval, line, arrow, and rectangles (see Figure 2.9). Putting object-structured graphics into group use, there arises the problem of concurrency control which is not significant in pixel-based group graphics. In a session of collaborative drawing, it is likely that two or more designers intend simultaneously to manipulate the same object appearing on the shared drawing surface.

To coordinate users' potential concurrent manipulations, the message-sending mechanism of object-oriented programming has been used in attempting to sequence concurrent processes at different sites. A good example is the design of object-structured group graphics in *GroupDraw* (Greenberg et al. 1991) (see Figure 2.10). This example shows an interesting attempt to integrate the design of communication primitives with the design of the graphics primitives. Greenberg's team built two instance variables into the root object of all graphics primitives: *ownerProcess* and *couplingStatus*.[10] By indicating who the owner of the process is, this variable serves to arbitrate contention in manipulating an instantiated object; by indicating the coupling

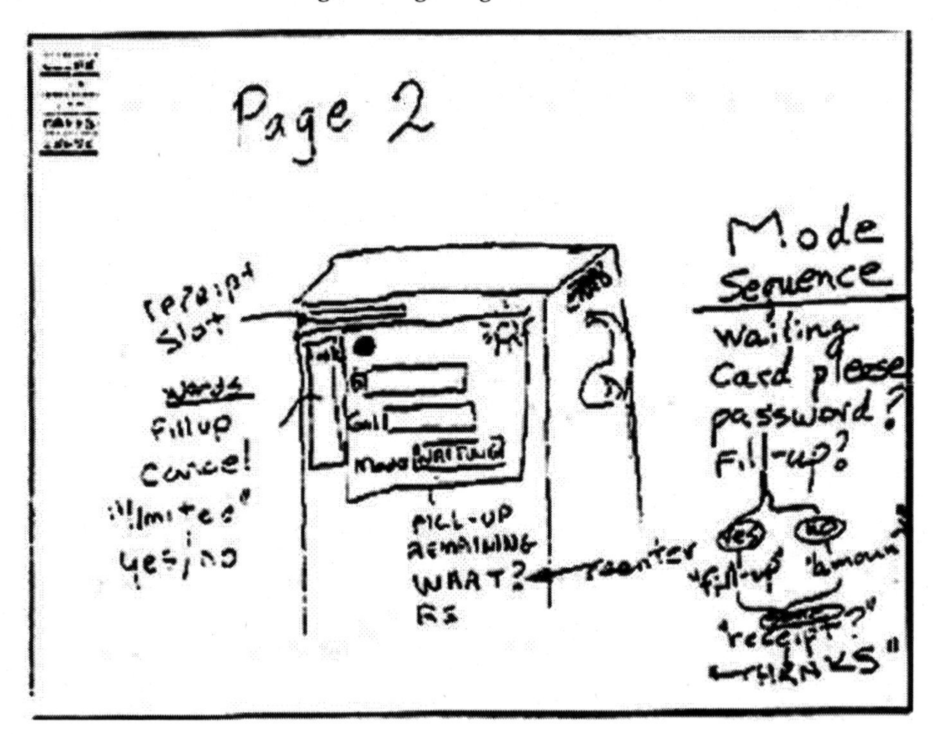

***Figure 2.7** The drawing/writing surface of the Commune workstation is comprised of transparent digitising tablets with styli; each digitizer tablet continually reports the position of its local stylus to the processor, and each user's stylus is represented on the screen as a pencil-shaped cursor producing marks in a distinct colour (from Minneman and Bly 1991).*

status of an object being "private," "public,"or "sharable," this variable determines the extent to which graphical objects are shared (Greenberg et al. 1991).

4 *Knowledge-based graphics.* In contrast to manipulating objects at a syntactical level (as in object-structured systems), graphical objects are defined and manipulated semantically in knowledge-based graphics. In a knowledge-based approach, graphics primitives are programmed in terms of abstract construction and operation components that are specific to particular design domains. Evaluations, explanations, advice, alerts, or criticism of graphical expressions constructed in those components can then be computed and presented to the designer at work. A shared drawing space connected to a knowledge base provides a set of domain-specific graphic constructs as a common design language shared by its user groups. The drawing operations, which enable a user's direct manipulations of objects, are more conceptually bound to the system's knowledge domain; for example, parts of a construction can be manipulated by changing the values of attributes associated with the graphical

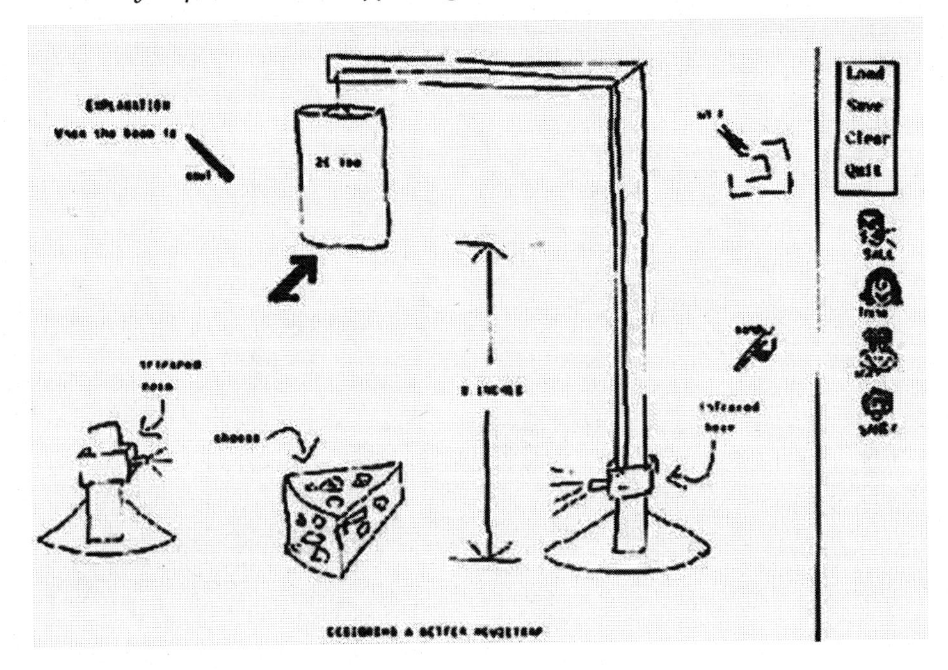

***Figure 2.8.** In GroupSketch, multiple mouse-driven cursors represented by different icons are used to convey participants' physical gestures such as drawing, typing, pointing, erasing, and directing public attention. The positions and movements of the cursors at all sites are visible to all participants in real time (from Greenberg et al. 1991).*

constructs embedded. At a higher level, a user may gain multiple views of a design by switching from one underlying construction kit to another.[11] The design of *XNETWORK* (a recent update of *NETWORK-HYDRA*) environment is such an example (see Figure 2.11), and it proposed a way of sharing drawing space through (indirect) collaborative construction of the common knowledge repository (Fischer et al. 1992; Reeves and Shipman 1992).

5 *Semi-structured graphics.* The term "semi-structured" refers to a picture representation resulting from a mixture of unstructured graphics (freehand or pixel-based) with structured graphics (object-structured or knowledge-based). Back in the early 1990s, there appeared two ways of enabling the use of semi-structured graphics in shared drawing space:
   - Formal drawings embedded in unstructured conversational sketches - the graphic editor vmac[12] plus visual languages for cooperation is an example of this (Lakin 1990) (see Figure 2.12).
   - Video captured images of freehand sketches superimposed on formal drawings generated by some graphics package - the design of *TeamWorkStation* presented a

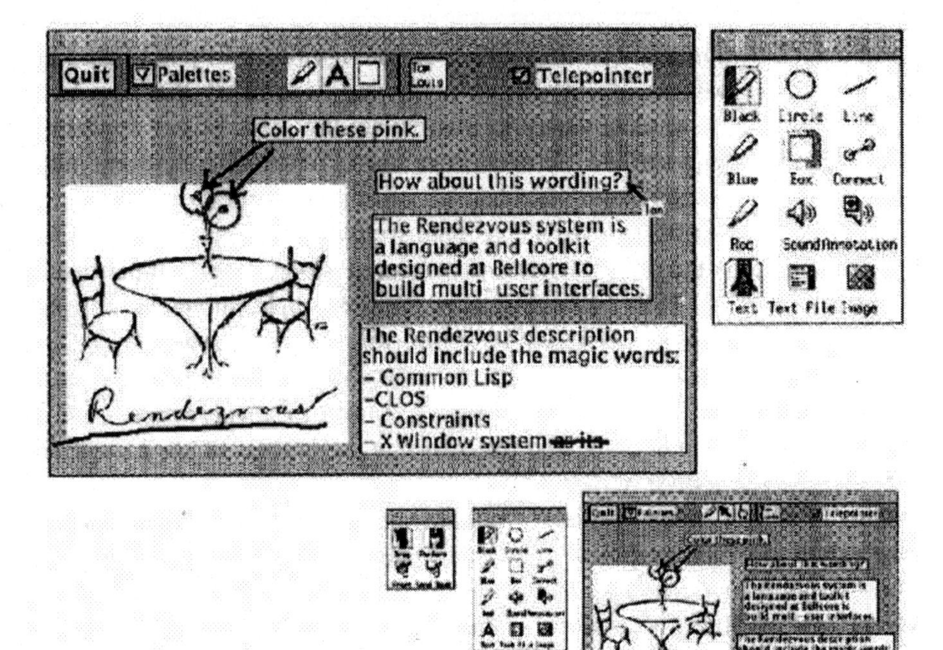

Figure 2.9 The Conversation Board developed at Bellcore provided a number of structured objects including oval, line, arrow and rectangle; after objects are placed on the shared canvas they can still be edited and moved (note that the "Rendezvous" sketch shown here was an imported image, which was originally drawn by hand and digitised, from Brinck and Gomez 1992).

shared drawing space where transparent video images containing hand-drawn expressions were overlaid with formal drawings constructed on a computer screen (Ishii and Miyake) (see Figure 2.13).

Communication networks and interprocess communications

As shown by these observational studies, the different perspectives have led to various choices of what communication networks are appropriate for supporting the various patterns of shared drawing events. The term "communication networks" has to cover a wider scope of system architectures in implementing shared drawing spaces; computer networks, as usually thought of, may not necessarily be involved here. Video networks, for example, have been exploited to support real-time collaborative drawing sessions held between remote working sites. When computer networks are used to serve the communication infrastructure of collaborative drawing surfaces, there arise issues of

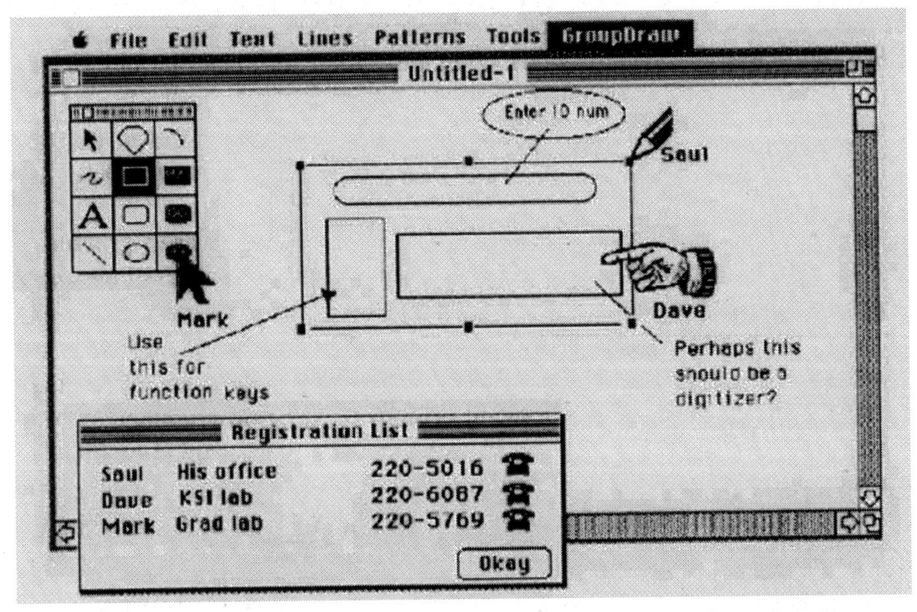

***Figure 2.10** GroupDraw was one of the first shared drawing systems using object-structured graphics to address the issue of concurrency control in collaborative drawing space (from Greenberg et al. 1991).*

concurrency control and maintaining consistency in data and view sharing. Message passing seems to be the most widely employed mechanism to handle interprocess communication which receives inputs from multiple users' drawing acts and delivers the computed end results to each participating site. It is possible to classify the network configurations in shared drawing spaces into four different types as described below.

1 *Hard-wired configuration.* This is the network design adopted by most video-based, or computer and video fusion approaches. The components of the network are purpose-built to perform the specific system functions as a shared drawing surface. In a fully video-based configuration, video cameras, monitors, and projectors are hard-wired for imaging, transmitting, and projecting the images of participants and the state of their work. There may be two main reasons why a hard-wired architecture is constructed: (a) to investigate how tele-presence can be realistically supported, which is conditioned by whether the configuration can convey cooperative work together with participants' *body language* (e.g., hand gestures, facial expressions, eye contacts, etc.) in the course of a remote meeting; (b) to simulate the elements of a natural setting of freehand sketching[13].

2 *Centralised configuration.* A shared drawing tool with a centralised communication structure can be explained by the "star" network topology, in which all workstations are connected via a single link to a central switching node (Sloman and Kramer 1987). Within this configuration, a central server runs a single application and conference

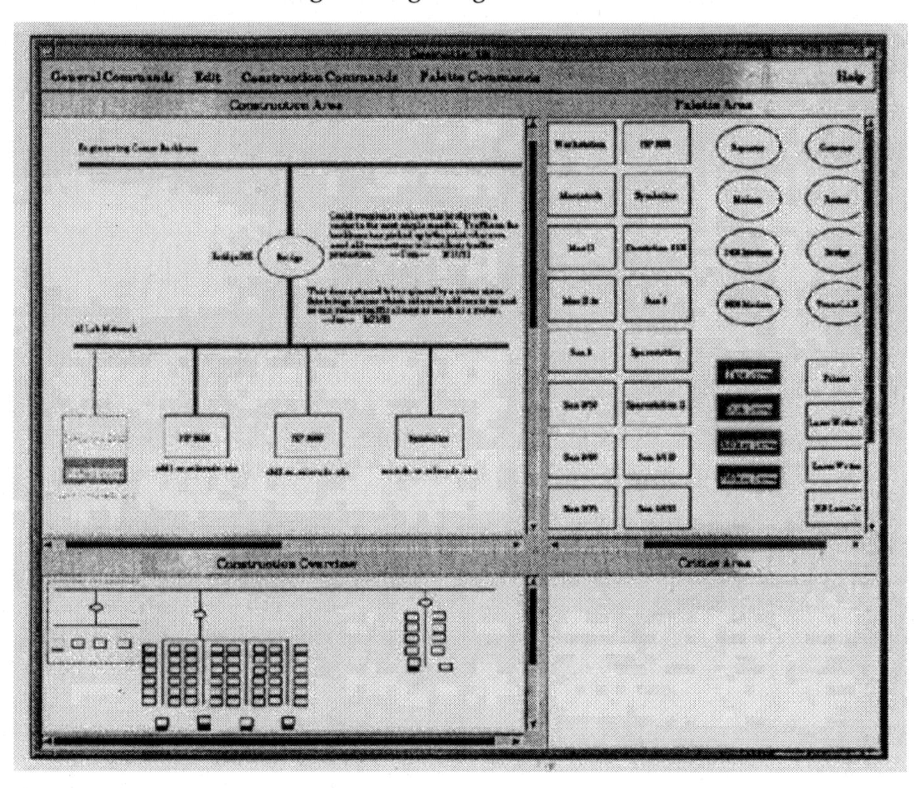

***Figure 2.11** The Construction Kit of XNETWORK allows for a connection between graphical construction and a knowledge base representing "group memory" of network design (from Reeves and Shipman 1992).*

process. The conference process handles most of the synchronisation and serialisation issues, the application process computes output of drawing functions from input multiplexed by the conference process. Each user's workstation runs a participant process (a user interface client), providing low level interactive graphics primitives.[14] Two constructs of interprocess communication in the client-server model have been imported from distributed programming: *Rendezvous* and *Remote Procedure Call* (RPC)[15]. The advantage of a centralised architecture is the relative ease of maintaining synchronous display among distributed participant sites. Apart from heavy network demands and being less robust in the face of network and host machine failure, a major problem of the centralised approach is that it does not support participants' sharing the processes of creating and using drawing expressions, which is an important requirement as specified in (Bly 1988; Tang 1991), since transmissions take place only after drawing acts are completed.

3 *Replicated configuration*. In contrast to the centralised approach, a replicated

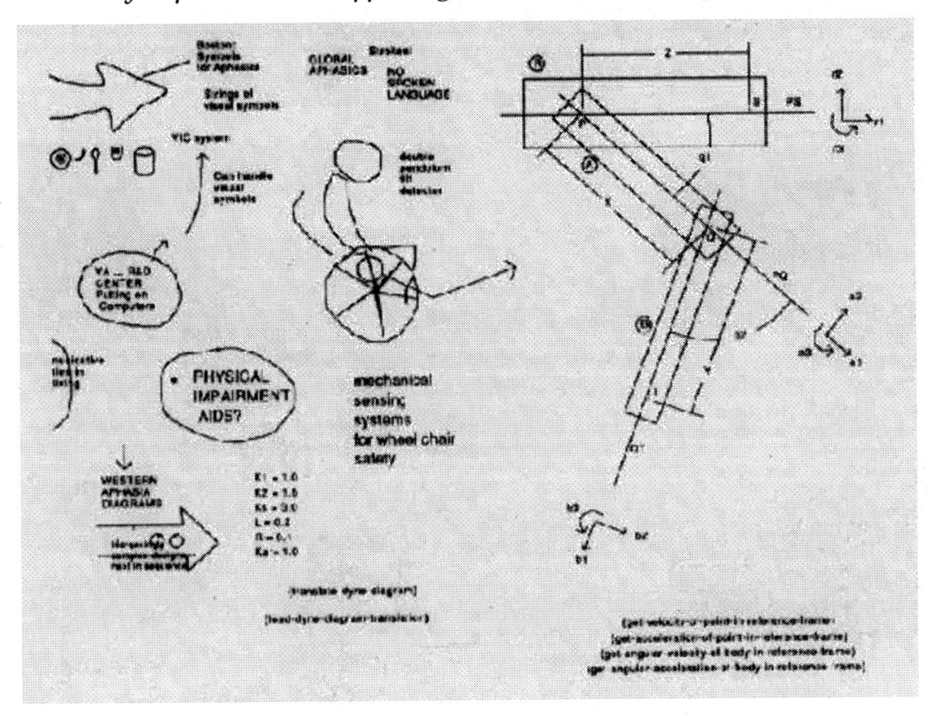

Figure 2.12 *Semi-structured graphics in the shared drawing space of vmacs enable both unstructured conversational expressions and visual language expressions which, as characterised by Lakin, are like "coagulated lumps in oatmeal." (from Lakin 1990).*

architecture runs a copy of the conferencing and application processes on every workstation that a user can interact with. The conference process at each site sends input to (and receives from) other networked sites, and passes received input to the application process for generating output and updating its resident display. Besides the merit of reduced network demands and, hence, gaining lower latency of application response on each local site, the replicated approach supports the conveying of drawing actions in the course of group meeting. A replicated configuration was built upon the serial bus or highway network typology, in which simultaneous transmission by multiple stations may result in interference; therefore, a media access control mechanism was needed to prevent or resolve contention for the transmission medium (Sloman and Kramer 1987). As a consequence, it is harder for a replicated architecture to achieve integrity of shared drawing surfaces crossing all participating sites.

4 *Hybrid configuration.* Here, a participant process run on every workstation may use a central conferencing process only for serialisation and synchronisation, and all other

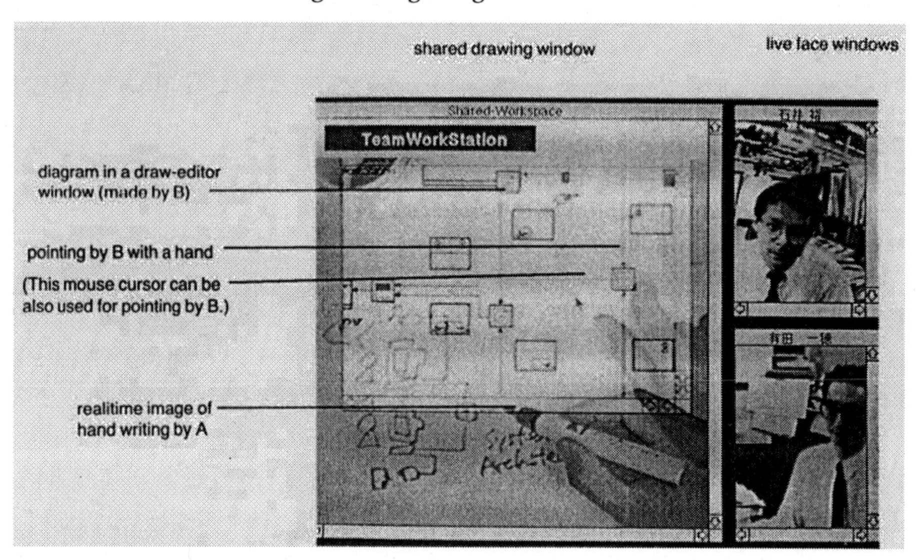

Figure 2.13 The shared drawing space of TeamWorkStation facilitates a translucent overlay of a desktop image of a freehand sketch on an image of a formal drawing, which was displayed on the shared screen of all participants for remote meetings (from Ishii and Miyake 1991).

shared drawing space acts are communicated directly between participant processes. Taking CaveDraw as an example, a hybrid configuration may consist of a central communication server that mediates the participating workstations which all run a copy of the application program (Lu 1992) (Figure 2.14). According to the CaveDraw experience, there remained the problem of concurrency because of the seemingly unavoidable time discrepancy between passing local events (by any one participant process) to the central communication manager and updating local displays (by the central manager).

Information storage and retrieval

To complete a shared design project, it may take participants hours, days, or even years to carry out individual as well as group tasks. The third cluster of design issues in shared drawing space involves the facilities to record and to manage the history of collaboration. Drawings and other forms of information created and used in the past may need to be brought back to the present for individual as well as group purposes. Several functions and interfaces of storing and retrieving information for supporting group design activity have been attempted.

1 *Camera or video record.* To make records of work results from collaborative drawing sessions, fully video-based workspaces often resort to video-recording and/or picture-taking. Since the video-based approach puts great emphasis on the sharing of drawing actions rather than on data sharing, storage and retrieval have not been

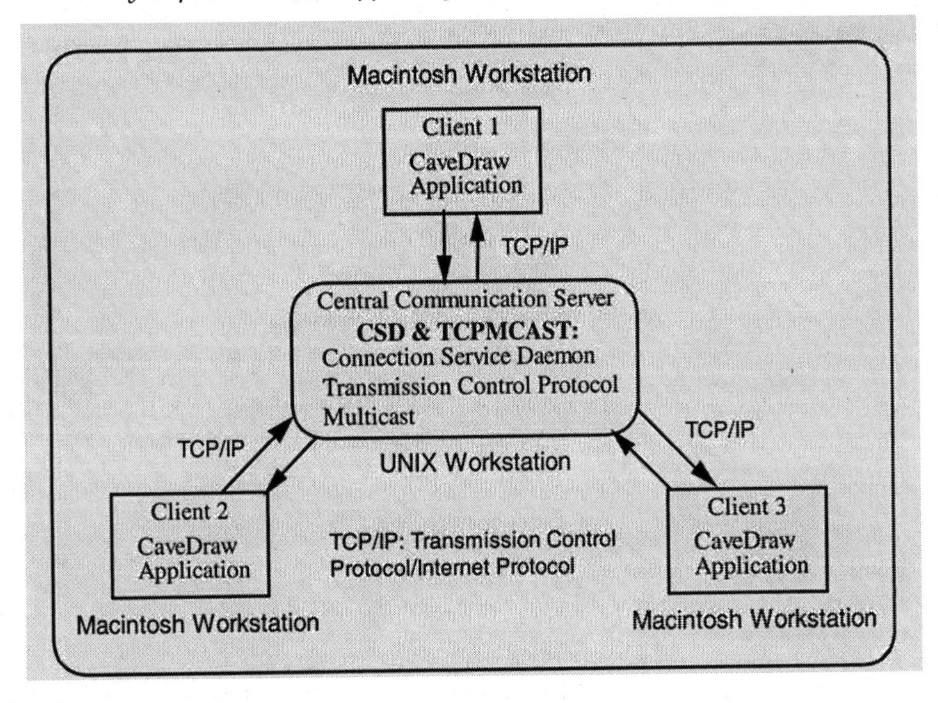

Figure 2.14 *The communication structure of CaveDraw—an example of a hybrid network configuration. (After Fig. 17 of Lu 1992)*

considered an essential function of a shared drawing surface. This decision is based on two explicit design rationales: that "drawings as artefacts in themselves are often meaningless" (Tang 1989); and that "hand gestures help participants to store (to remember) design discourse" (Tang 1991).

2 *Workstation drawing files.* Workstation-based group drawing tools can make direct use of the file management facilities of most operating systems. In supporting storing and retrieving data during collaborative drawing and design sessions, there appeared to be different metaphors of storage and retrieval which lead to interfaces with various design features:
  - *(Time-stamped) pages or scenes.* In systems like *GroupSketch, Commune, vmacs,* or *We-Met,* the interface for storage/retrieval simulates the pages of a note pad or flip chart. New "pages" or "scenes" can be continuously issued for making marks; and, when facilitated by a history mechanism, one or more pages can be reloaded onto the current working surface.
  - *(Time-stamped) miniature sheets.* In *Boardnoter,* sketches are saved into reduced views as a collection of miniature sheets visible on the common screen. When retrieved, each sheet can be re-displayed at full screen size (Stefik et al. 1987).

- *Drawing layers.* In *CaveDraw*, the use of drawing files by a team of designers comes close to that of tracing paper by a design team, which facilitates transparent overlay drafting (Woods and Powell 1987). Drawings sketched on one or more layers can be stored in a single file; when recalled, it can be displayed together with other resident layers on one's drawing surface.
- *Catalogue items.* In *NETWORK-HYDRA*, design representation was saved in a catalogue serving as a repository of designs constructed by participants over, perhaps, a long period of time. The catalogue contains several items, dealing with different aspects of the design task such as graphical construction, design rationales, and design specifications. Existing items in the catalogue can be accessed and copied into a new construction by modifying what is retrieved; and completed instances of design can be archived into the catalogue for future use or reference (Fischer et al. 1992).
- *Content-directed retrieval.* As one of the visual languages designed for cooperation, Lakin proposed a novel function of retrieving drawing files by their contents. It was suggested that content-directed retrieval might proceed by having users write and draw their expressions in a formal visual language, *TEXT-GRAPHIC-QUERY*[16] (Lakin 1990).

Multi-user interfaces

In supporting multi-party synchronous graphics interaction between collocated or remotely separated designers, there arise several novel design features, which are not common in traditional single-user drawing systems.

1. *Telepointer.* A telepointer is a large cursor that appears on a common screen of a meeting room or on every workstation connected over a local network. As an interface device for group interaction, it is designed to be manipulated by participants (one at a time) to point to specific locations on a shared drawing surface. Telepointing was often said to simulate the conveying of information by hand gesturing in group drawing activity. Limited effects of telepointing have been reported in, for example, *Boardnoter* and *Conversation Board* (see Figures 2.5 and 2.10 respectively).
2. *Multiple cursors.* The rationale underlying the design of multiple cursors is that multiple identities can be attributed to local cursors used by individual participants. Identity of a cursor can be defined by, for instance, the colour it produces, the name of its (current) user, or the gesture indicator it serves (i.e., pen, marker, eraser, pointer, etc.). But there is a trade-off between the support for the various modes of gesturing and the support for rapid switching among drawing, writing, and other actions (Bly 1988; Brinck and Gomez 1992; Greenberg et al. 1991).
3. *Group vs. individual views.* A shared drawing space may allow users to have different views that are local to individuals. On this issue, there appear different approaches to user-controllable view sharing:
   - A view-sharing facility based on the "What You See Is What I See" (WYSIWIS) principle serves all participants sharing strictly the same view during collaborative sessions; events taking place on any one site immediately affect the current states

of the shared drawing surfaces appeared on other sites. In this case, no individual views are allowed for private work.

- The WYSIWIS principle is relaxed to some extent so that the effects of manipulating parts of a shared drawing space can be kept locally. A participant can turn away from a public domain by scrolling the shared drawing interface to different places (e.g., *in We-Met*, see Figure 2.6; and in *GroupDraw*, see Figure 2.9), or by moving onto another drawing layer (e.g., in *CaveDraw*, see Figure 2.15).
- A participant's workspace consists of an individual screen and a shared screen. With this workspace set-up, a user is given great flexibility of controlling what and when things are to be made individual or public. Given the view sharing design of *TeamWorkStation*, a participant can simply drag a drawing or a document image on his or her own individual screen onto the neighbouring screen which can be synchronously shared by all networked working sites (see Figure 2.16).

Other dialogue facilities

In supporting rich and complex collaborative design activity, communication channels other than a shared drawing surface are often provided in parallel. At least three other dialogue channels have been regularly used to complement the design of shared drawing space:

- Audio links are most widely employed to support participants talking about parts of sketches while drawing or pointing at them.
- Video links are used not only to capture desktop images but also to transmit facial images of participants. There is a difference in the extent to which captured facial images are integrated with a shared drawing surface, and these are separate (e.g., in

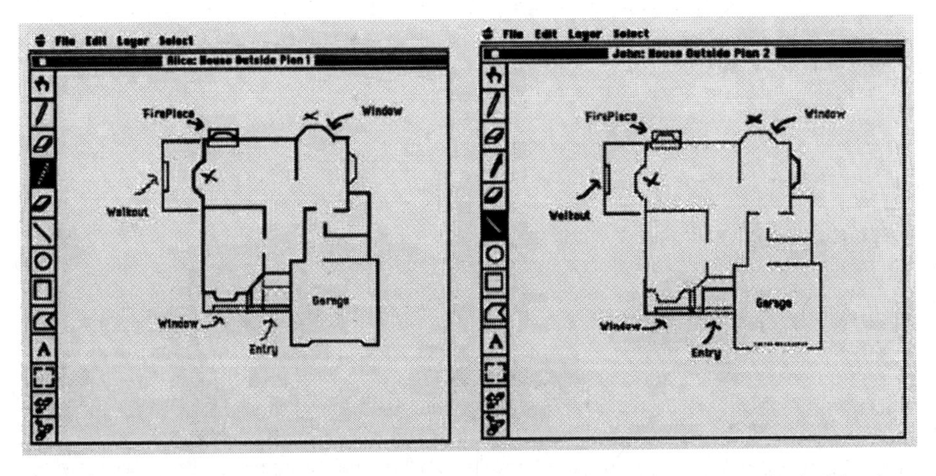

Figure 2.15 The overlapping layered approach in CaveDraw: Two participants can have individual drawing surfaces when co-working on different layers (from Lu 1992).

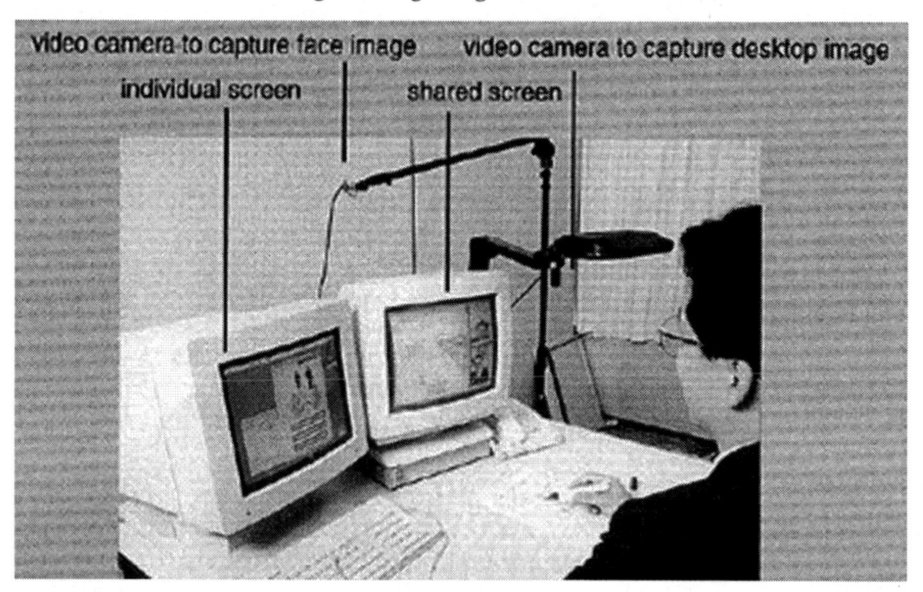

Figure 2.16 The juxtaposition of individual screen and shared screen in TeamWorkStation (from Ishii and Miyake 1991).

VideoDraw, Commune), juxtaposed (e.g., in *TeamWorkStation*), and entirely fused (e.g., in *ClearBoard*).

- A messaging service is provided in connection with a shared drawing surface, enabling users to send textual or graphics messages in the course of collaboration. The *MUSK* system was designed with a text-based messaging service (Crampton 1987); the visual language *VISUAL-MAIL*[17] working together with *vmacs* enables users to mail text-graphic pages (Lakin 1990). A mailing service is a practical alternative channel when audio/video links are not made available, or collaborative work is not based on synchronous interaction.

2.4 Group uses of shared drawing spaces

In the above, I have reviewed the empirical or conceptual studies of group drawing activity and the approaches to developing prototypes of shared drawing support tools. To interrelate the two parts of the survey, this section presents a discussion of the uses of the prototype tools developed. Seen in a broader view, it seems that the research and development of shared drawing support tools have aimed to serve three different group uses of shared drawing tools: conversation, management, and meeting. Each use points to a way of collaborative working and, thus, a set of goals for system support. This differentiation leads to a grouping of the prototype developments into three categories:

1. *As conversation media*. As seen, most of the prototype designs are concerned with supporting real-time drawing interaction among remotely located participants who,

presumably, come from more or less the same professional or technical background. To be used by group members, the functions of shared drawing space are devised mainly for supporting conversations. Systems built for this purpose may employ graphics primitives and drawing operations ranging from freehand sketches to object-structured graphics. Since facilitating tele-presence is a major goal, the provision of multi-modal dialogue channels is as important as the set-up of a shared drawing surface. Concerning shared use of the data resulting from conversations, however, conversation-oriented systems provided little or no functionality for storing and retrieving during collaborative work sessions.

2 *As management media.* In the attempts to support collaboration via graphical and/or knowledge representations of design ideas or rationales, a shared drawing system may be characterised as a medium for users to manage collaborative design work. Two different views of supporting users' management behaviours have been expressed. In the case of *CaveDraw*, one addresses the need for direct communication among participants; this view considers the users' need to organise design ideas that are often represented in graphical form. The other addresses indirect and long-term collaboration, considering the need of representing design knowledge in textual form; the formally captured knowledge is further connected to the system's drawing space. The shared drawing surface, as in the case of *XNETWORK*, is shared in the sense that the graphics construction space has formal (internal) communication with the state of a shared knowledge base containing design rationales written down by, perhaps, designers involved in the past.

3 *As performing media.* The third design perspective considers shared drawing space activity taking place in meetings, participated by two working sub-groups; namely, a single or multiple performers (or demonstrators, facilitators), and several viewers (or commentators). Group drawing or diagramming activities normally take place within a meeting room equipped with a conferencing system, as in the case of *Boardnoter*, where the performer has exclusive access to the large common screen, making presentations to other participants during a brainstorming session. In *We-Met*, the strict WYSWIS feature is relaxed to the extent where participants can scroll a shared scene into private areas without affecting each other, but the change of a whole scene remains exclusively controlled by one individual. Concerning the choice of drawing primitives, the use of semi-structured graphics in *vmacs* is a novel idea, which seeks a compromise between the support of agility/openness by conversational graphics and that of expression processibility by computerised interpretations.

Based on the review of the system features and the grouping of the prototypes, an overview of the survey is presented in Table 2.3. Given the fact that the research on shared drawing support tools is a continuously growing field, the above categorisation may not be applicable to later system research and developments. Nevertheless, this survey of the early experiments with collaborative drawing tools indicates the scope of research regarding human factors and technical designs that have been explored by the researchers. Looking back, these early initiatives can be considered as the origin of computer supported collaborative design.

2.5 Shared drawing spaces of the Pre-Internet age: A Retrospect

In this chapter, drawing on a selection of observational studies and research prototypes, I have presented a survey of the early research and developments in shared drawing spaces that took place during 1987 through 1993. Immediately after this period, we started to see the arrival of the Internet and the World Wide Web on a global scale. The Internet and Web-based network technologies introduce a whole new horizon of computing communications. Given that the new generation of computer network infrastructure is now widely available, implementing dedicated computer networks for the purpose of creating shared drawing or writing spaces is no longer a primary issue of concern to the developers of collaborative systems. Developments in CSCW and CSCD have since moved onto exploring more innovative and versatile multi-user interface and group coordination mechanisms. I shall show some examples of the more recent developments in Chapters 7 and 8. But based on the review presented in this chapter, I may draw some general concluding remarks on the enterprise of understanding collaborative design and developing system architectures and interfaces for shared drawing space. The early basic research into the nature of human communications involving graphical interaction has been worthwhile and the lessons learned from the various projects have been valuable to later system research and developments. In my current survey of the early experiments in supporting collaborative drawing and design, distinctions between *drawing* and *designing* have not been made explicitly. Clearly, there are differences in system objectives between, say, simple electronic sketch pads for people to make free-hand drawings together, and more formal/structural systems for designers to develop and manage design tasks cooperatively through networked graphical media and interface. Drawing and designing cover somewhat different aspects of human cognitive and communicative activity, which may therefore give rise to the diverse recognitions of what is to be supported by computing communications.

Drawing can be considered as a general form of human communication in which pictures and texts are the traces left by communicative acts, which took place sometime in history. Shared perception and understanding of these traces, thus communication, is highly conditioned by the sharing of *drawing actions* that produce and use the physical traces.

Designing, on the other hand, seems to centre around some previously or currently existing, or yet to exist, design artefacts, which does not always involve drawing activity; structures of design artefacts can be introduced by or emerge from designers' *design thinking*, which often conditions communication and collaboration among co-working designers.

Systems that support group drawing activity must deal with the issues of supporting direct communications among users who may not be of face-to-face presence. Unstructured graphics prove to be a good choice for a shared drawing surface responsive to demands for (a) the intimacy between what is drawn and the actions of drawing, (b) the speed needed to maintain conversations, and (c) the freedom of making expressions. Structured graphics has been attempted in group drawing mainly for increasing the functionality of drawing surfaces and for concurrency control; but its real effectiveness in supporting synchronous graphics communications remains to be demonstrated.

SYSTEM FEATURES \ RESEARCH PROTO-TYPES		Conversational Mediums								Management Mediums		Performance Mediums		
		Video-White-Board	Com-mune	Team-Work-Station	Group-Sketch	X-sketch	Group-Draw	Clear-Board-2	C'nver-sation Board	Cave-Draw	NET-WORK-HYDRA	Board-noter	vmacs Visual Langs.	We-Met
Graphics Primitives & Drawing Operations	direct drawing	●	●	●	○	○	○	○	○	○	○	○	○	●
	pixel-based	○	●	●	●	○	○	●	○	●	○	●	●	●
	object-structured	○	○	●	○	●	●	●	●	○	○	○	○	○
	knowledge-based	○	○	●	○	○	○	○	○	○	●	○	●	○
	semi-structured	○	○	●	○	○	○	○	●	○	○	○	●	○
Communication Networks	hard-wired	●	●	●	○	○	○	●	○	○	○	●	●	○
	centralised	○	●	○	●	●	○	○	●	●	●	●	●	○
	replicated	○	○	○	●	●	●	●	○	●	○	○	○	●
	hybrid	○	●	●	○	●	○	○	○	●	○	●	●	○
Information Storage and Retrieval	cameras/videos	●	●	●	○	○	○	●	○	○	○	○	○	○
workstation drawing files	miniature	○	○	○	○	○	○	●	○	○	○	●	○	○
	pages	○	●	●	●	●	●	○	●	○	○	●	●	●
	layers	○	○	●	○	○	○	●	○	●	●	○	○	○
	catalog	○	○	○	○	○	○	○	○	○	●	○	●	○
	patterns	○	○	○	○	○	○	○	○	○	○	○	●	○
Multi-User Interfaces Design	telepointer	○	●	●	○	●	○	○	●	●	○	●	○	○
	multiple cursors	○	●	●	●	○	●	●	○	●	○	●	○	○
	WYSIWIS	●	●	○	●	●	○	○	○	○	○	●	●	○
	relaxed WYSWIS	○	○	○	○	○	●	●	●	●	○	○	○	●
	separate screens	○	○	●	○	○	●	○	○	○	○	●	●	●
Other Dialog Channels	audio links	●	●	●	○	○	○	●	●	○	○	○	○	○
	video links	●	●	●	○	○	○	●	●	○	○	○	○	○
	messaging	○	○	○	○	○	○	○	○	○	●	○	●	○

LEGEND: ○ Not Developed or Used ● Developed or Used to a Degree ● Developed or Used

Table 2.3 A categorisation of the research on shared drawing support tools in terms of different group uses and system features.

Systems that support group work in design need to provide shared drawing space for construction and communication. The problem of how to integrate formal or informal representations of graphical objects and design ideas with an understanding of communication and coordination in design remains to be explored more deeply. We have seen two alternative approaches to group design support systems: structured construction with knowledge representation and semi-structured conversation with formal visual languages. Yet, regarding the former, we have not seen approaches that support group design processes based upon synchronous and heterogeneous representations and uses of design knowledge. In respect of the latter, the concept of supporting autonomy in addition to that of heterogeneity may need to be further addressed such that (a) organisationally, there is less discrepancy between individuals making contributions and gaining benefits or satisfaction through the use of technologies, and (b) administratively, teamwork can be freed from one single managerial ambit. As a general guideline, further development of computer supported collaborative design environments should always take into account what is essential to design practice that demands better support for interdisciplinary collaborative processes as well as integrated design products.

Drawing, being important to design in many fields, is a natural mode of communication among members of a design team, and the research pursued in the design and implementation of shared drawing space has indeed been worldwide. It can be said that the understanding and developments made, or yet to be made, in this research area can make important contributions to the advancement of CSCW systems. Though research into computer-supported collaborative drawing or design poses distinctive system design issues and technical concerns, it is useful to see how the specialisation may be related to the issues raised by the CSCW research community in general. CSCW research has a general concern: the nature and aspects of supporting *group processes*. Among many others, Paul Wilson has outlined the aspects of group processes basic to the theory, practice, and design of CSCW (Wilson 1991). By referring to, in particular, the first two aspects suggested by Wilson, I would like to point out that research in collaborative drawing and design presents interesting issues that are worth further investigating:

1 ***Individual work patterns.*** The design of group support tools and working practices has to take into account individual work habits and predilections. In this aspect, design activities present potentially a high degree of idiosyncrasies among group members. How can users specify personal constructs or tools as prerequisites for supporting individual work patterns? This consideration opens up potential system design issues concerning (a) participants deriving individual design spaces from the design space they share, and (b) how a common design space may emerge and evolve from the interaction between individual design spaces.
2 ***Representation of organisational knowledge.*** Research on structures of graphics can be further pursued to develop the representation of organisational knowledge graphically (see, for example, [Star 1989]). Can a simple graphical approach contribute to a better management of organisational knowledge, which by its nature tends to be difficult to locate, recall, and update?

The above issues, seen in this survey and in relation to a broader agenda of CSCW research, are of particular interest to my enquiry into the requirements for computer support in collaborative design. Especially, these issues are closely related to the view of design as modelling complex objects explained in the previous chapter. Collaborative design seen in this perspective involves a group of designers who communicate and coordinate with each other in the processes of modelling design objects. Drawing activity will be considered as a part of designing activity, which involves other kinds of activity that may be better understood and supported as a modelling activity. In the next four chapters, I shall introduce and discuss cases that illustrate design as modelling and collaborative design as teamwork in design modelling. This perspective points to representation and communication requirements that have not been fully addressed by the research and experiments in supporting collaborative drawing as reviewed in this chapter.

Summary

Aspects of shared drawing space activities were observed by a number of CSCW research groups to identify specific opportunities for developing support tools. There are four main aspects identified in this survey: the four spatio-temporal patterns of group drawing events, the exchange of action- or representation-oriented information, the uses of homogeneous or heterogeneous drawing/design tools, and the distinction between group and individual ownership. By focusing on the various findings of group drawing activities, research prototypes have been built by the various research groups in different countries during 1987 through 1993, i.e., prior to the rapid growth of adopting the Internet and World Wide Web on a global scale. This period of early studies of group drawing and design activities and building research prototypes can be considered as the origin of computer-supported collaborative design.

It is shown that there were five clusters of design issues emerging from this survey of 11 prototype systems. Under each issue, a number of technical solutions have been developed and tested in field trials. First of all, the choices of drawing primitives, ranging from natural free-hand strokes to highly structured graphical construction, seem to reflect how the researchers see drawings realised in group work contexts. Among them, the approach of *semi-structured graphics* shows some interesting results in supporting dynamics and heterogeneity occurring in teamwork.

Collaborative drawing can take place in a geographically (and temporally) distributed manner, if each participant's drawing platform is connected through some architecture of communication networks. Given drawing primitives and a network, information produced by participants is currently stored in and retrieved from either videotape or a workstation filing facility. Computer-based group drawing tools require some unconventional user interface designs, including a telepointer shared by all users, multiple cursors associated with each user, concurrence control over graphical objects, and the separation between private and public drawing surfaces. Most prototype tools seen in the survey were incorporated with other dialogue channels, including audio links, video links, and message sending over networked drawing spaces.

When the early prototypes were experimentally put to use, we observed three categories of group uses of the early collaborative drawing tools. Tools in the first category, with the largest number of system implementations, were able to function as media of real-time graphical conversations. Through graphical and/or knowledge representation of design ideas or rationales, tools in the second category served as media for managing group design work. Systems in the third category were used mainly by a single performer communicating with other collaborators in a meeting room.

3 Collaborative Design and Discovery of Metaphors

> A true team between architect, painter and sculptor, aiming at an organic synthesis of their work by symbolic association, needs an intensive exchange by its members. For only a thorough, mutual penetration into the ideas of the various team-mates may lead to that kinship of spirit which can make a creative entity out of the individual contributions. ……
>
> If contributions of sculptors and painters are desired by a client, he should let the architect choose congenial collaborators at the initial designing stage of a building. Only such foresight will enable the architect to build up his team in time to let each member take part in the creative process of the basic space-conception, a process so essential to the result of final unity in the building.
>
> — *Walter Gropius, 1961*[18]

Having surveyed the early CSCW experiments in developing collaborative drawing systems, in this and the next three chapters, I shall turn to design studies on collaborative design. The common approach to design studies adopted in these chapters was to conduct analyses of *case histories* as a way of searching for the evidence of team working in architectural design. Following the notion of design as modelling complex objects introduced in Chapter 1, I use the evidence from the case histories to illustrate collaborative design activity in terms of *collaborative modelling of complex objects*. In doing so, I continue to formulate some abstract descriptions of *teamwork patterns* of collaborative design from the case histories, and it is from these standpoints I explain the basic requirements for computer supports for team working in design. It has to be said, there is always the danger of making generalisation on the basis of limited case studies. My intension here is simply to present some similarities among these case histories studied, which seem to suggest some discernable teamwork patterns in terms of flows of information. I shall therefore ask readers to be open-minded regarding if these descriptive patterns or models of collaborative design can be applied to elsewhere. Furthermore, I show that the design studies presented here leads to a formulation of the requirements for computer-based collaborative design environments, but it is inherently problematic to lay down any specific footprints upon which complete practical systems can be built accordingly. Given the long and rich history of architectural design, it seems to me that the best we can do is to build up a kind of *requirements library*, with which we can collect user requirements over time as references to the design and testing of collaborative design software components.

This chapter presents an observation of how designers work together to achieve integrated building design on the basis of individuals' design contributions. Based on what can be seen from these case histories of building design, I show that one of the keys to fruitful design collaboration lies in the discovery of *common images* resulting from inter-personal communication, and *common design metaphors* emerging from group interpretation of the common images can function as a communicative device that enables participants to collaborate effectively in the processes of design development. A set of *constraints on collaboration* is identified via a situation-theoretical analysis of the *bottom-up* scenario of collaborative design. I shall then discuss the implications for building collaboration-supporting tools with respect to the constraints derived.

3.1 Studying how designers work as teams

Like many other human activities, the design and construction of built environments has always involved many individuals working jointly as teams. One simple reason is that buildings or more broadly the built environments are often too complex to be handled thoroughly in all aspects by a single person. Building design has been like this in the history of design as portrayed by Khan (Kahn 1935). Modern design practice for the built environments is even more of the case as described by Michael Middleton (Middleton 1967). This seems to run into the opposite of the impression that we often have: famous buildings are the products of famous architects. In many ways, the false impression has been led by contemporary commercial marketing machinery. The combination of buildings with their architects' images is a powerful advertising strategy that is often adopted to capture potential buyers' or the general public's imagination. In reality, successful buildings are seldom the outcomes of some omnipotent architects who work alone; more often than not, architects may act as project leaders, but good buildings are the outcomes of months, if not years, of collaboration among designers of different expertise working in various domains of building design.

Certainly, collaborative design or group design process is not an entirely new subject to the research community of Design Studies. If we look for research literature in this subject matter, we can see a variety of interpretative frameworks that have been proposed by design researchers. One of the most widely used metaphors of group design process is that of design as *game*. Lawson reviewed several design games that were specially devised to simulate group dynamics in architectural and urban design (Lawson 1990). Working under the game metaphor, Habraken and Gross constructed a computer program called "concept design game" that can record and then replay sessions of interaction involving two or more participants playing the game of control distribution and territorial organisation (Habraken and Gross 1988).

Regarding research methodology in general, protocol analysis and case study are alternative major methods used by design researchers. I shall mention two examples. Based on his analysis of a documented design dialogue between an architecture student and a studio master, Donald Schön proposed a theory of "reflection-in-action," acting as an epistemological framework for design learning and teaching (Schön 1985; 1991). A studio-based empirical study of a group design process participated by a group of seven architecture students was carried out by Tony Ward (1987), in which the students went through group processes and developed *archetypes* for a commercial complex building scheme; the archetypes were mainly recognised by gathering together a collection of physical models made of cardboard by the participants.

It is useful to point out the contrast between design theorists and CSCW researchers as far as design studies is concerned. As we have seen in the preceding chapter, understanding collaborative design in CSCW research is not so much about the contents of design. System developers seem to be more interested in the generic activities of drawing and design communication including the subtle body languages observed. On the contrary, design researchers often talk about their findings of group processes with close references to the contents of design tasks undertook by the participants. Obviously, different research issues will lead to different types of system development. As it is, most collaborative drawing tools are generic in the sense that they are not content specific; whatever users draw is irrelevant to how the systems function. But, we may ask, are there any benefits if we introduce some of the elements derived from the contents of design studies to the making of computing tools? Will architectural designers find them more useful to their tasks at hand when working with computing communications systems that reflect some verified understanding of collaborative building design? These are questions with no straightforward answers. Designers seem to need both open generic as well as content specific tools under different circumstances.

Still a few more words need to be said about the nature of the case studies presented in this chapter (and also Chapter 5). In comparison with the examples of design studies mentioned above, again I take a different stand. I consider that an investigation of cooperative design can be founded by analysing, not simulated nor controlled, but naturally created and evolved design expressions taken from historical cases of building design. I have explained the basic rationale of doing so in Section 1.6 of Chapter 1. As will be shown by the end of this book, my current study of the case histories has drawn some conclusions not seen in other research literature.

First of all, among the case histories presented, we shall see different kinds of design expressions that correspond to (a) the participation of multiple individual design worlds, and (b) the production of integrated designs. Secondly, there exist relations among the different types of design representation; it is by explaining the relations that I arrive at some conceptual accounts of group interaction in design modelling. What is important in grasping these concepts is that they appear not bound to particular structures of design products or processes nor to specific strategies of design organisation or management. This study comes up with some descriptive concepts that call for further elaboration on generic patterns of teamwork in design.

The remainder of the chapter is organised as follows. A total of five case histories of building and landscaping design is introduced in the next section. Following the case studies, Section 3.3 puts forward an abstract of what I call a *bottom-up* scenario of collaborative design. As an analytical framework for deriving the constraints on design collaboration, a brief introduction to the situation-theoretical account of information established by Barwise and Perry is given in Section 3.4. Finally, an exposition of the teamwork pattern derived from the case histories is presented.

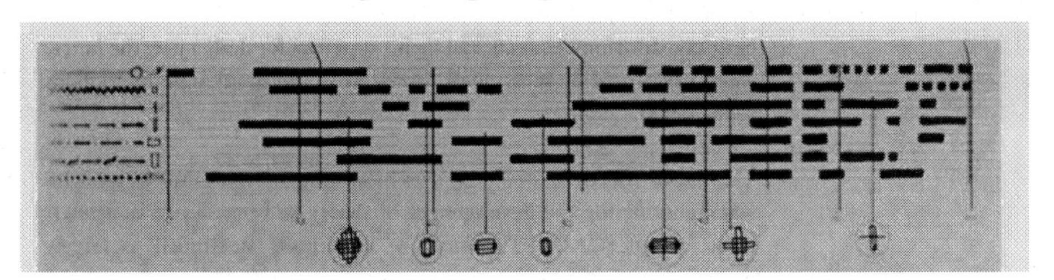

Figure 3.1 The landscape designers' constructing and manipulating score as a way to model fountain patterns and actions over a period of time (from Halprin 1969, 56).

3.2 Communicating space-conception in design: some case histories

(Case 1) Between score and diagram

The first case history is of an urban landscape design project. The project was about the design of the Seattle Center Fountain in Seattle, USA, during 1962 through 1964. The design team consisted of (a) landscape architects (Lawrence Halprin and Curtis Schreier), whose main responsibilities were to design the waterscape of the fountain, and (b) a mechanical engineer (Daniel Yanow), who was an expert on fountain engineering (Halprin 1969). By examining the drawings and diagrams published, I consider the collaboration between the landscape architects and the fountain engineer can be described in terms of the following.

The scoring and diagramming spaces. The landscape architects used a particular representation scheme called "score" for modelling fountain patterns and actions in a temporal frame (Figure 3.1). A score has two dimensions: one for regulating multiple temporal sequences, represented in certain lengths of bars; the other for configuring spatial structures of different fountain stages (platforms), represented as point, square cross, rectangle, etc. By manipulating the bars, a score reveals different compositions of active fountain stages against the inactive ones over a period of time.

The mechanical engineer used "diagrams" to model mechanical components for piping, jetting, and sprinkling design (Figure 3.2). A pool piping grid was composed in a system of graphical symbols corresponding to a set of design objects whose attributes were specified in words and numbers. In relation to the piping grid, a mechanical section was constructed to convey sectional information. Following that a correspondence was established between the mechanical components and the graphical symbols, the mechanical engineer could change his configuration of the attributes and relations of particular design objects by manipulating parts of the diagrams.

A common space for projecting water effects. Figure 3.3 shows a series of graphical expressions of *squiggles* spreading over a regular *grid*. This evidence, it seems to me, suggests that a common modelling space shared by the landscape architects and the fountain engineer was being used, combining the designs in the score and in the diagram, by which a sequence of water effects could be *projected* in some graphical

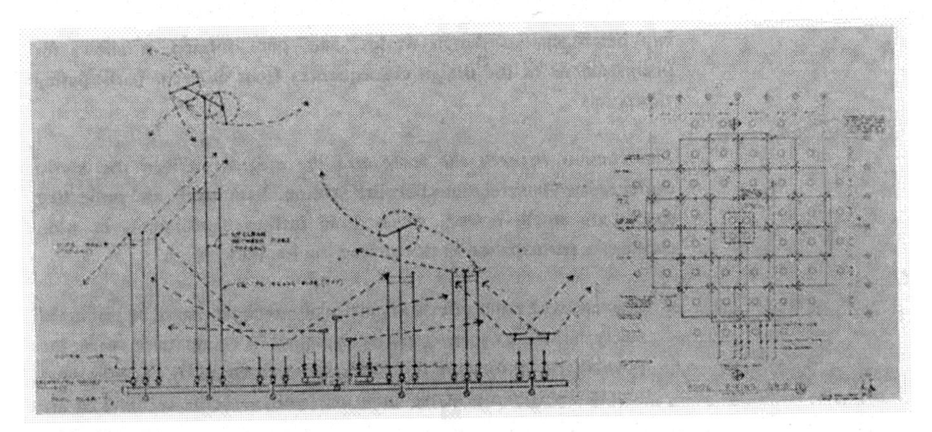

Figure 3.2 The mechanical engineer's constructing and manipulating the piping diagrams as a way to model the behaviours of the mechanical components (from Halprin 1969, 56).

symbols. I think what we see here is an example of a set of *common images* generated, allowing for *interpretations* of the design consequences from various viewpoints. It is clear that the images of water effects can be interpreted both in the landscape architects' view (the actions of fountain stages as scored over a time span), and in the mechanical engineer's view (the fountain kinematics concerning the motions in pipes, jet heads, and sprinklers, as configured in the piping grid and mechanical section).

The interrelations between the score and the diagram. Given the above evidence, two interrelations between scoring, diagramming, and projecting spaces are worth noting, which yield further explanations of what constitutes participation in developing the fountain design:

- Sequences of water effects at particular moments cannot be projected solely in the landscape architects' scoring space nor in the fountain engineer's diagramming space; the ability of projecting these effects is conditioned by knowing what fountain stages are active and what mechanical devices are operating on those active stages, plus how they will behave—a convergence between two individual modelling spaces whenever a projection is undertaken.
- Modelling actions performed in individual spaces change not only the state of score or diagram but also the state of the common image when projected; the mechanical engineer may take further actions upon what he interprets as changing water effects propagated from the landscape architects' actions in changing the score, and vice versa—communication and coordination are called for to resolve disagreements or conflicts thus arising.

(Case 2) "The fishbone layout"

The second case history draws from an environmental design for a large-scale industrial

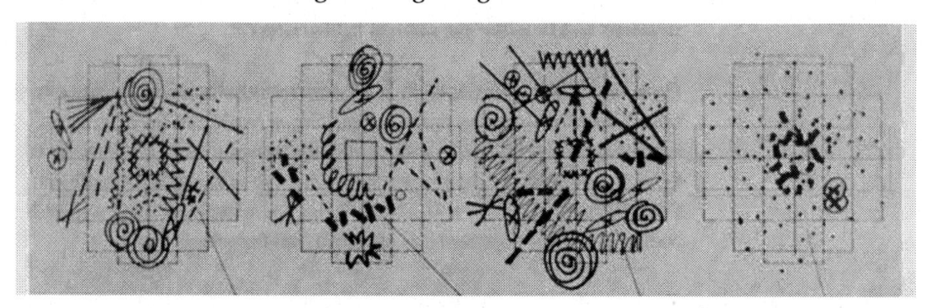

***Figure 3.3** The graphical indications of a shared fountain modelling formed by a combination of the landscape architects' scoring and the mechanical engineer's diagramming, which can project water effects, allowing for different interpretations (from Halprin 1969, 56).*

building. The Cummins Research and Engineering Center in Indiana, USA, required a major upgrade to "rearrange its service ductwork to the structure and to baffle the indirect light sources." (Lam 1977, 125-129). During 1964 through 1968, a project team was formed to design the re-servicing scheme. There were (a) structural engineers from The Engineers Collaborative, (b) lighting engineers from William Lam Associates, and (c) mechanical engineers from Cosentini Associates. According to Lam, the project was developed jointly by the participating engineers' overlay diagramming, which can be described in terms of the following:

Distributed diagramming spaces. Each engineering discipline developed design in its own domain-specific diagramming space. We may say that there were at least three domain-oriented diagramming spaces involved in the project: structural, mechanical, and lighting engineering (Figure 3.4). The designers employed special coding systems to represent the modelled building components in each of the diagramming spaces.

Collaborative environmental design through overlaying. According to Lam's recollection of the design process, a "fishbone layout" was evolved from the group working, which proved to be economical and satisfactory to all three aspects of engineering requirements (Lam 1977 see Figure 3.5).

Collaboration through overlay diagramming. Supposedly, the emergence of the fishbone image was a result from the participating designers *overlaying* their domain-specific diagrams (perhaps, repeatedly) in the design process. But an important point is that the resultant image in the name as well as in the form of "fishbone," provided the members of the design team with a shared conceptual entity that allows for individual interpretations regarding how detailed design of building parts should be further developed.

In Case 1, we saw an example of two participating design worlds interrelated with each other by the projection of end design results (the waterscapes of the fountain). The

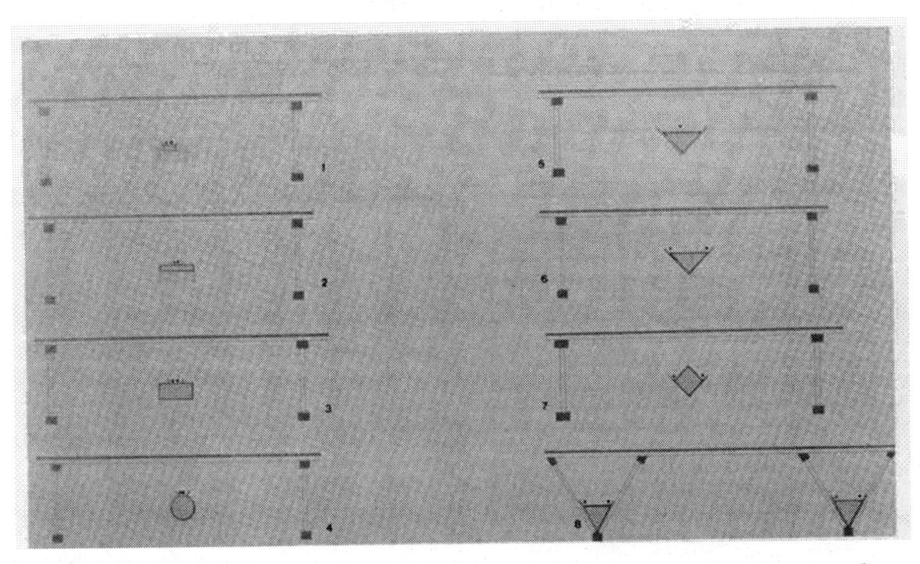

***Figure 3.4** Multiple diagramming spaces in different layers showing the participants' heterogeneous coding systems for modelling aspects of the building design (from Lam 1977, 126).*

current case, however, shows an example that participants can further *articulate* a projected common image into various parts that bring up more elaborate definitions of parts of the design (the further differentiation of "fishbone" in to the spinal cord and ribs, in this example). Designers working in various areas then take newly differentiated parts of as design references for further development of domain-specific schemes. Given that common images can be evolutionary (i.e., the designers evolve their ideas about what the "fishbone" will look like and how it will function), we should recognise that the process of individual articulation is an important part of team working. It is likely that parts of an existing common image get identified and developed by different individuals, which, in turn, further contribute to a state of the common image. Teamwork in design is therefore maintained by a *to and from* relation between the individual and the common image.

The method of overlay diagramming and drafting has been developed for sometime in design practice. For instance, Woods and Powell have documented the method and recommended it as a standard team practice (Woods and Powell 1987). In general, the practice of overlay diagramming can facilitate design collaboration in terms of the following:

- *Overlay diagram construction.* A participant can construct diagrams on top of extracted common images which may contain parts of diagrams drawn by other designers working in different areas.
- *Overlay design checking.* A design can be evaluated jointly by designers checking

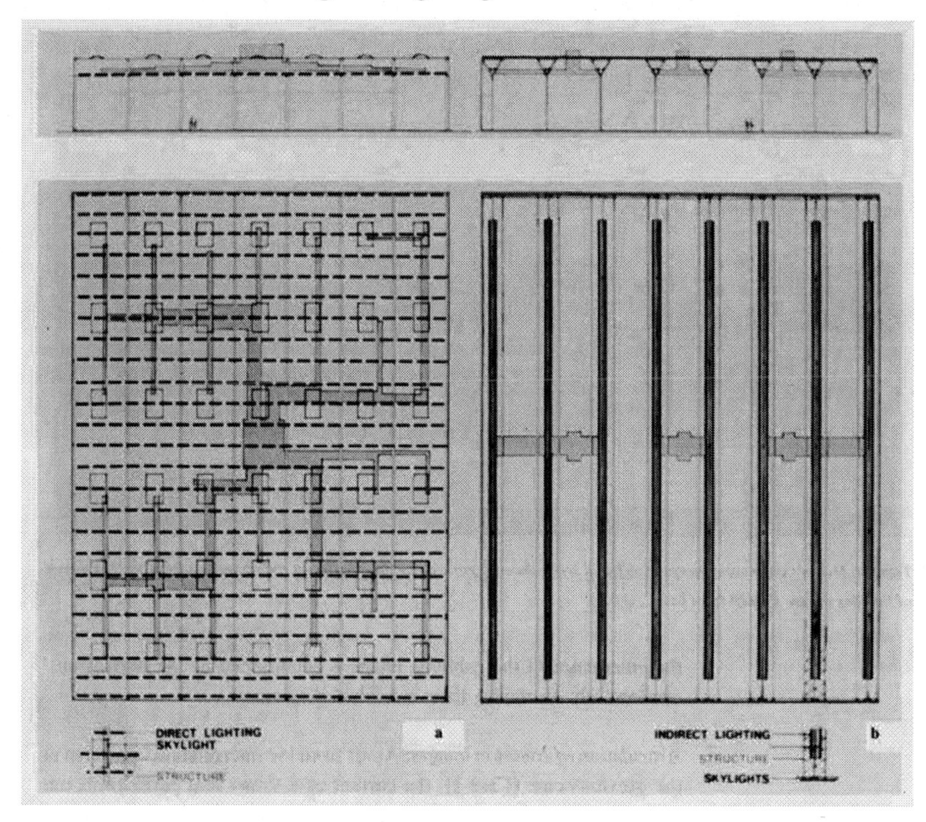

***Figure 3.5** The combined images of structural, mechanical, lighting design solutions evolved from the participants overlaying domain-specific diagramming (from Lam 1977).*

overlaid consequences according to certain design criteria such as detection of spatial clashes.

- *Overlay design amendment.* A participant can modify parts of diagrams by referring to the diagrams underlay for various purposes (e.g., geometrical, structural, or aesthetic); and one designer's amendments may cause other changes to be looked at by others.

(Case 3) "An architecture dreaming about fish"

The Elephant Group of Team Zoo was founded in 1971, consisting of members from U-Ken, an architectural design office led by Professor Takamasa Yoshizaka (once an assistant of Le Corbusier), and members of his urban development staff at Wasada

University, Japan. Team Zoo was commissioned by a client to build a private house in Kamakura, Kanagawa Prefecture (Speidel 1991, 15 and 32):

> The Domo Serakanto, built in 1974, is a house christened 'coelacanth' because of its (planar) shape, created within a long, heterogeneous group process: five members of the design team designed individual sections, which was allowed to show in the clear joint within the complete building. They gain unity from the image of fish.
>
> It took the owner two years to find this place. They took us to the site in the summer of 1972. What came out from almost three years of struggle was a coelacanth that crawled out from the sea. Domo Serakanto, with its gill, spine, horn, cilium, teeth, antennae and scales, it appeared in the wind and sank in the light. Domo Serakanto is a fish dreaming about architecture, an architecture dreaming about fish.

As a German architect working closely with Team Zoo, Manfred Speidel commented on the essence of team working developed by Team Zoo over years (Speidel 1991, 15):

> Team Zoo had discovered that teamwork, even when markedly individual, can still produce a coherent whole, if a common image exists that each individual can interpret differently, and that allows a great deal of scope.

Why the image of coelacanth? I should say that the image of the "fishbone" in Case 2 and the image of the "coelacanth" in the current one are both associated with "fish" is purely a coincidence. No explanations were given by the original authors and designers as why it was the image of fishbone that guided the design team to develop a satisfactory organisation of the building's new environmental service system. However, as an observer, I find no difficulty in recognising that the plan shown in Figure 3.5 indeed looks like fishbone. And I believe that most people would agree with the association unless they never saw fish before. In the case of the Domo Serakanto project, it is probably easier for us to appreciate the analogy of fish with the design of a house due to the building site's close proximity to the sea. But why did the members of Team Zoo converge on a common image as specific and unique as coelacanth in designing the house? Again, I did not find explanations for this in the original document. From a research point of view, I do not consider that the specific contents of common images emerged or discovered in collaborative design processes are that important. The "fishbone" and "coelacanth" images are extremely important to the design teams who were engaged in those projects at those times. But the images are *internal* to the design teams while working on those two projects. It seems to me that the finding and use of coelacanth as an analogy in the house design was operated by Team Zoo on a *one-off* basis as I do not find reappearance of coelacanth images in their other building projects.

(Case 4) An image of "vessel"

A new chapel for the Fitzwilliam College at the University of Cambridge was built during June 1989 through February 1991. Design of the chapel was developed by a team

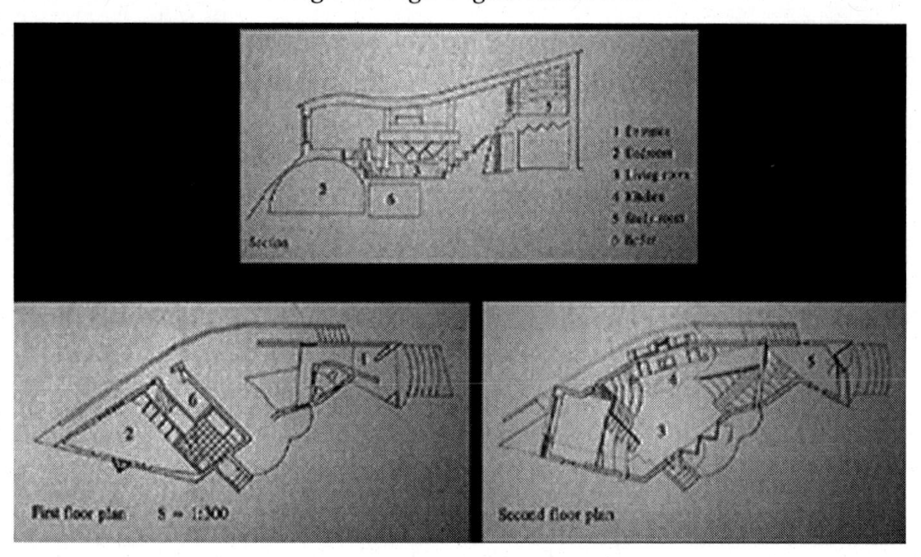

***Figure* 3.6** *Plans and section of Domo Serakanto, Team Zoo, 1974 (from Speidel 1991).*

of designers at the MacCormac Jamieson Prichard Architects (Blundell Jones 1992). One of the salient features of Fitzwilliam Chapel is its strong archetypal image of *boat* or *vessel* (Figure 3.7). When interviewed later by Lawson regarding the design processes, Richard MacCormac reflected on how the concept and image of *vessel* came along that enabled the design team to establish greater clarity and coherence among parts of the building (Lawson 1994, 64-69):

> At some stage the thing became round but I can't quite remember how. Eventually the upper floor began to float free of the structure supporting it. The congregational space became a sort of ship. However, it was not until quite detailed problems were considered such as the resolution of balcony and staircase handrails that this idea was fully understood and the 'vessel' took on its final shape and relationship to the supporting structure.

The dynamism between parts and wholes. The discovery of the "vessel" in the chapel design was certainly a breakthrough to the design team as it provided the team with an overall architectural order and strategy that informed many design moves to follow. However, the image of vessel was fluid in a way that its final form was subject to the articulations or resolution of components of the vessel such as the balcony and staircase's handrails. Clearly, design collaboration does not end at discovering some strong images that members of a design team can feel associated with. The case above shows just an example of a continuous interaction between design developments of parts and an evolving common image as a design whole. Another point can be made about Case 4 is that, unlike the preceding three cases in which the emergence of common images took

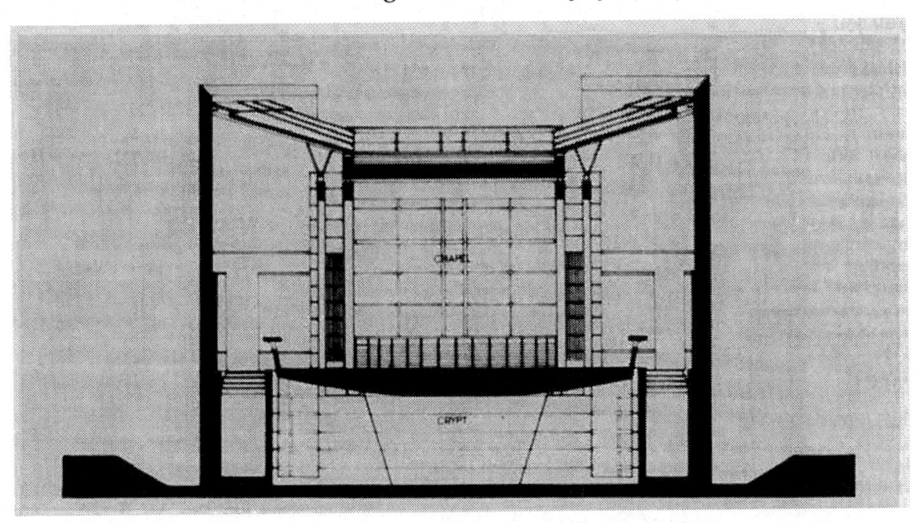

***Figure 3.7** An image of vessel: A section of the Fitzwilliam College at the University of Cambridge (from Lawson 1994).*

place at a planar view from top, the design team made a sectional discovery. And in the next case study, we shall see a case of discovering common images that is not based on a planar or sectional view but a three-dimensional vision.

(Case 5) Images of "oculus" and "wall"

The fifth case history takes a look at another institutional project undertaken by the MacCormac Jamieson Prichard Architects who were commissioned by the Cable and Wireless Plc to design its Headquarters and Training Building in Coventry, England. Again, when interviewed by Lawson, MacCormac recalled vividly a session in which he and one of his assistants were working on the residential part of the project (Figure 3.8) (Lawson 1994, 60-64):

> At the very beginning of the process the centre of the scheme was a circular courtyard, but later I thought this was wrong. By then we have this 'V' idea going in which the building opens out in a 'V' shape rather like the wings of a bird ... towards this wonderful landscape. Then suddenly I had this idea that the courtyard should be pulled into an oculus, a sort of eye shape which would reflect the dynamic of the whole project.
>
> I can't quite remember what happened and either Dorian or I said it's a wall, it's not just a lot of little houses, it's a great wall 200 metres long and three storeys high ... we'll make a high wall then we'll punch the residential elements through that wall as a series of glazed bays which come through and stand on legs ...

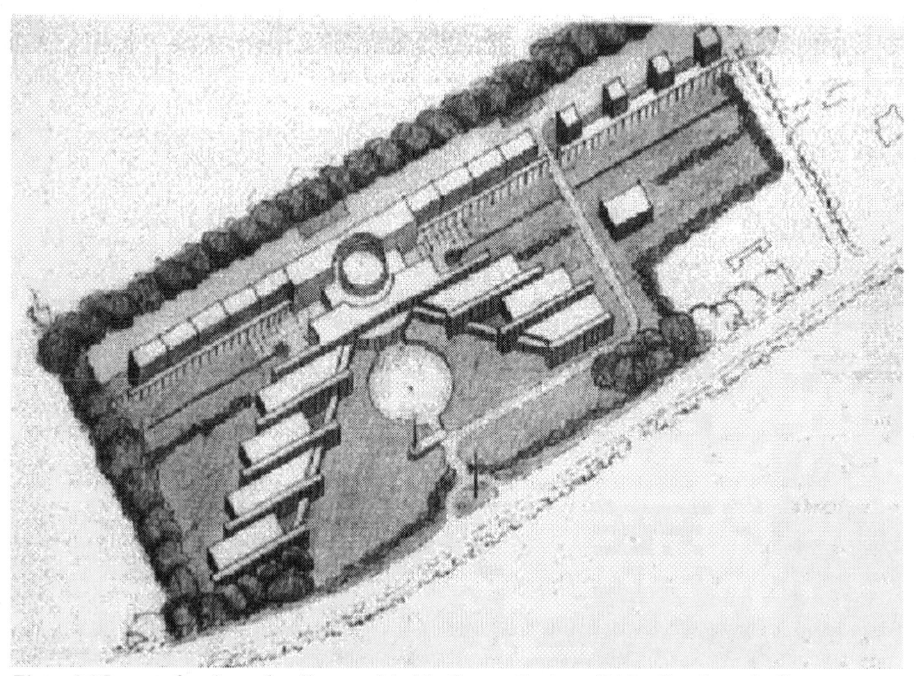

Figure 3.8 Images of oculus and wall emerged in MacCormac Jamieson Prichard's scheme for the Headquarters and Training Building for Cable and Wireless (from Lawson 1994).

Naming of common images. According to Lawson, the naming of an architectural concept or image is a cognitive advice for designers to better understand and communicate what is emerging. Once named in a certain way by some member of the design team, the concepts and images can then be shared by the rest of the team; the team then work out the implications embedded in the concept by testing its appropriateness or potentials of developing various parts. It also serves to keep the design team together, as it were, in "ensemble." (Lawson 1994, 60). MacCormac recalled the following:

> I or somebody else comes up with an originating idea, some idea that seems powerful enough to generate a scheme and to subsume a lot of decisions within it ... it needs somebody in the team to pick up the ball and run with it. I find that I seize on somebody in the team who understands what the crisis is ... you have to find this person who sees what it is about otherwise it's hopeless.

The above review of case histories is by no means exhaustive; nor do I claim that collaborative design by discovery of common images is the most typical (or, representative for that matter) teamwork pattern that design researchers or practitioners may agree or identify with. In fact, I will discuss a different pattern of collaborative building design in later chapters based on another set of case histories. Nonetheless, in the next section, I intend to give a descriptive theory of collaborative building design processes with regard to the case histories discussed.

3.3 Shared discovery of metaphors: A bottom-up scenario

Design in general is a *messy* process with no clear markers that we can point to as a beginning and end moment of the process. To propose a general scenario of collaborative design, some form of abstraction is required. In what follows, I describe *bottom-up* scenario of teamwork in building design that involves the discovering and sharing of common images or metaphors. This abstract intends to layout the terms for the main constituents of the team working processes observed from the case histories. A more detailed exposition based on a situation-theoretical framework will follow in the end of the chapter.

A bottom-up scenario

At the inception of a design project, a team of designers of different schools of expertise or perspectives firstly set up individual workspaces with which the designers generate and manipulate design expressions meaningful to specific design aspects (domains). Designers' setting up individual workspaces may be distributed over several remote working sites, and they may carry out initial design work addressing domain-specific requirements in a parallel manner. At some later stage, the designers decide to join their individually developed design schemes together in a shared workspace. To resolve issues of making satisfactory connections among individual design parts, team members embark on design integration tasks in the common workspace. Typical design integration tasks may involve the following:

Aggregation—achieving design composition on the basis of putting together spatially parts of the design expressions made by individuals into larger wholes; or

Projection—achieving design composition on the basis of projecting overall design consequences by combining functionally the design properties modelled in the individual design expressions.

Going through further sessions of design meeting, some kind of images or concepts may be discovered or recognised by some members of the project team, conveying the cognition that the integrated design *looks like* certain things or *functions like* certain ways. The common images revealed are then shared by other team members for domain-related design interpretations. The common images may be given certain names with which the participants can make references to the images or concepts when conversing with one another. This is a stage of collaborative design at which team-shared *design metaphors* emerge.

Design meetings and the emergence of common design metaphors may give rise to new problems or opportunities awaiting domain-specific design deliberation in individual workspaces. There can be two sources of making design moves: (a) new parts are needed following the attempts to aggregate individual contributions, and (b) new parts or changes in existing parts are identified because of team members' interpretation of individual local schemes in the light of the common images emerged. Participants may be thus motivated or requested to carry out further design developments whereby they may modify or articulate their local design schemes previously arrived at.

The design team may convene further sessions of team meeting to review newer states of local design developments. If necessary, members of the design team may carry out further aggregation, projection and interpretation, which may lead to revised relationships between local design parts and the common images evolved. Many interactive sessions may occur between domain design developments and design team meetings until all project participants are satisfied with the design results both in terms of the parts and the whole.

The term "bottom-up" is used here as a characterisation of the teamwork pattern as collaborative design is mainly driven by the initiatives from local design developments—that is, designers create various "parts" to start with, and then by putting those parts together, a "whole" is gradually built up collaboratively. If we consider the above abstract scenario of collaborative design reflecting properly the case histories seen earlier, several questions can be raised regarding the group interaction or dynamics in collaborative design:

- How can common design images be aggregated or projected collectively by participants if they know little about each other's design domain? What is required for participating designers to understand each other such that a sensible adding up of individual contributions can take place?
- What can be seen on the surface is that aggregation of design parts (or, projection of overall design consequences as in the Seattle Center Fountain project) can give rise to new developments in some (local) design domains, and vice versa, but can we give a clearer account of the communication involved between the integration taking place in a common workspace and the distributed domain-specific design developments in individual workspaces?
- Does the sharing of common images emergent from group processes play a role in the coordination of design changes made by team members among distributed individual workspaces? Is there a kind of group dynamism in collaborative design that can be further illustrated by revealing the interrelationships between the sharing of common images and distributed developments of design parts?

In exploring the questions raised above, I develop an exposition of the bottom-up scenario of collaborative design based on *a situation-theoretical* framework. The framework is also applied in my study of another teamwork pattern presented in Chapters 5 and 6. Before presenting the situation-theoretical analysis, a brief introduction to the situation theory is given in the next section, which provides an elaborate way of investigating the role of information flow in human communication.

3.4 A situation-theoretical framework of description

Jon Barwise and John Perry have been researchers at the Center for the Study of Language and Information (CSLI), Stanford University. In 1983, they first published their theoretical study of (formal) semantics of natural language (English) as Situation Theory (Barwise and Perry 1983). Since then the book has become one of the most influential studies on natural semantics and has been translated into German, Japanese, and Spanish. Barwise and Perry's situation theory mainly contends that meaning in natural language does not exist entirely within words and sentences but resides largely in the situations and the attitudes brought to it by speakers and listeners involved. As a reader of the theory of situations, I was particularly interested in their accounts on *information flow* and *constraints on information flow*. It seems to me with these situation-theoretical devices the fundamental communicative aspects of design collaboration can be articulated more fully, and a clearer and coherent description of the requirements for the design of computer support can follow.

Basic ideas of the theory of situations. A comprehensive introduction to Barwise and Perry's situation theory is not considered necessary here. Instead, it would be more useful to introduce some of the basic building blocks of the theory that are most relevant to my enquiry in collaborative design. Furthermore, the situation theory was originally intended as a development of mathematical foundation for formulating a formal semantics of natural language. Omitting the detailed mathematical structures of the theory, I shall focus rather on the intuitive insights brought up by the theoretical framework. It should be said that situation theory remains a theory of information that is under continuous development by many other researchers, especially, in the formalisation of its various aspects. The basic ideas of situation theory introduced below are drawn from the books written by Barwise and Perry (Barwise 1989; Barwise and Perry 1983; Devlin 1991).

Situations. A situation is a structured (limited) part of the world (concrete or abstract) discriminated (or individuated) by an agent. Information is always taken to be information about some situation, and we can have a proposition saying that some information is made factual by some situation, or, to put it another way, that some information is true of some situation.

Situation types. A situation type is an abstract (mathematical) object (in situation semantics) to represent real situations. Two unique situations belong to the same type if some "type abstraction" can be applied onto the two situations. To give an example, the situation where "Mary was running in Hyde Park at 3:30pm," and the situation where "John was running in Princes Street at 5:00pm" belong to the same type of situation in which "Someone is running in some location and at some time."

Information flow. One situation can contain or carry information about another situation only if there is a systematic relation that holds between the situation types that the two situations belong to respectively..[19] This corresponds to Dretske's "Xerox principle" (Dretske 1981, 57):

> If A carries the information that B, and B carries the information that C, then A carries the information that C.

Constraints. Constraints are the systematic relations between types of situations that allow one situation to contain information about another situation. To quote Barwise and Perry's original remarks on constraints (Barwise and Perry 1983, 51):

> ... for reality to support intelligent life it must be highly structured. What happens at one place and time *must* contain information about what has happened or will happen, elsewhere and elsewhen. So we need to provide an apparatus in the theory of situations to characterize this structure.

An agent's acquisition of information from a situation is circumscribed by those *constraints* of which the agent is aware, or to which the agent is attuned.[20] Situation theory characterises constraints by introducing a primitive relation between types of situations, the relation of *involving*. A constraint **C** can be expressed by a situation **S** involving another situation **S'**, by writing $C: S \Rightarrow S'$.

Logic of activity. The theoretical implication of the situation theory is that it provides an ontological foundation for analysing any natural activity (e.g., linguistic communication, mathematical reasoning, visualisation, etc.) in the world of human beings. That is, given the constructs and rules of the situation theory, we can describe the internal structure (i.e., the logic) of a natural activity. The logic is a logic of information, concerning the constraints on the flow of information. As Barwise put it (Barwise 1989, 52):

> When we search for the logic of some activity, what we are after is the collection of constraints ($S \Rightarrow S'$) that govern this activity. For example, the logic of perception consists of the set of constraints that govern perception.

A descriptive framework for collaborative design. There are several reasons why I think that an application of the situation-theoretical framework briefly introduced above in collaborative design is appropriate and useful:

Collaborative design is a natural human activity involving multiple (intelligent) agents. In principle, like any other natural activity, group design can be subject to a situation-theoretical treatment that will lead to an analysis of the constraints that govern collaborative design; and this seems to correspond to what I have been trying to derive from the design case histories.

The constructs proposed in the theory of situations, such as "situation types," "information flow" and "constraints," can provide a kind of quasi-formal framework for describing the communicative aspects of group interaction in collaborative design in a structural way.

There are precedents of applying situation theory in various areas that are relevant to design collaboration. For example, in enterprise integration (EI) modelling, Christopher Menzel and others argued that model integration should be based on the notion of information flow not of translation (Menzel, Mayer, and Sanders 1992); in analysing and describing the fundamental social structures that influence human communication behaviours, Keith Devlin used situation theory as a descriptive framework and developed what he called an "endogenous logic" (i.e., internal logic) for sociologists to work with (Devlin 1992a).[21]

However, in the course of attempting to apply some of the aspects of situation theory, I found it necessary to introduce design-related concepts to be included in defining the situation-theoretical types. In this way, the application appears more meaningful and useful. However, this is not to produce my own version of a situation-theoretical framework applicable in analysing group design activity, but to arrive at a degree of abstraction pertinent to further detailed discussion on the issues prompted in the case histories. I shall start with a definition of situation types in collaborative design as follows.

Situation types in collaborative design. To give a categorisation of situation types in collaborative design, some parameters of situation are introduced first according to the concepts of "uniformities," "indeterminates," in situation theory. I therefore propose that any situation of collaborative design can be characterised in three basic elements: modelling space, modelling act, and design state.

Modelling Space—An agent[22] (or, a number of agents) defines some *design constructs* in some medium, an abstract system, or even an existing modelling space with which a design expression can be constructed and manipulated. In the case of a single agent, we have *individual modelling space*, otherwise, *group modelling space*. The collection of constructs defines the scope of a modelling space.

Modelling Act—An agent (or, many agents) takes some action in the modelling space that (s)he or they have defined. The kinds of actions are to do with, for example, representing, constructing, transforming, or mapping etc., of constructs or instances of design expressions.

Design State—Whatever modelling actions taken by an agent individually or by a group of agents collectively will lead to a design state. In the language of situation theory, a design state is an *information carrier*, which contains information that may or may not have an overt representation. However, design information carried by a design state can be known to an agent or among agents through communication channels other than visual ones. A collection of design expressions made by an individual in his or her own modelling space, a designer's list of questions to be answered (regarding a piece of drawing, for instance), or some design effect known to a group of individuals are examples of design states. We then have a method of generating a range of design states for a scenario of collaborative design by

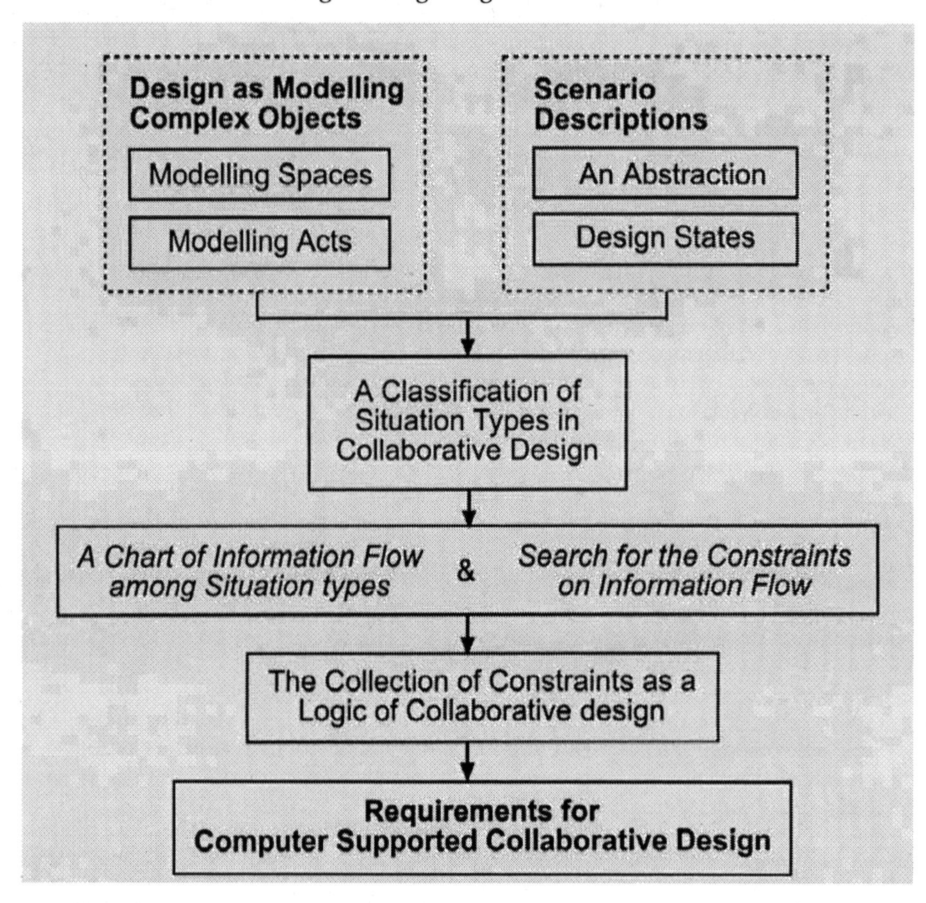

***Figure* 3.9** *An overview of the approach to the analysis and description of teamwork patterns in collaborative design.*

constructing *space-action* matrices; in operational terms, the matrices constructed prompt us to enumerate in a comprehensive manner the types of information carriers encountered in the group design processes.

By including the above three parameters into a triple of (*acts, spaces, states*), a definition of situation types more specific to the activity of collaborative design can be formulated as follows:

Situation Types ≡ **Modelling Acts** *located in* **Modelling Spaces** *leading to* **Design States**

Information flow among situation types. According to situation theory, if there are systematic relations between two situation types, then information carried by situations of the types can flow from one to another. In the case of collaborative design, we can now form a situation-theoretical view on the information flow among situation types in group design as defined above. In particular, it seems to me that we are in a better position to examine the flow of information among situations occurring in a group modelling space as well as in individual modelling spaces. A description of information flow of this nature, I believe, will shed some light on the web of group interaction in collaborative design.

Constraints on the flow of information. When (possible) information flow among situation types is identified, the things to search for next are the systematic relations (i.e., constraints) that allow the flow to take place. In our case, the way of finding the constraints is to infer the conditions under which one design state gives rise to another. Given the abstract of the bottom-up scenario described earlier on, the interrelations between the design states are relatively easier to be identified with.

Logic of collaborative design. In my situation-theoretical description of collaborative design, I aim to arrive at a set of constraints that are said to govern the flow of information among types of (group) design situation. According to the situation theorists, a set of constraints can reveal a "logic" of a particular kind of natural activity. Applying that premise to the study of collaborative design, I may say that I elicit a *logic of collaborative design* with respect to the bottom-up or any other kinds of constraints identified (we shall soon see a different scenario in chapters 5 and 6). But, we may ask, what is the use of the logic of design collaboration found? In my current approach, the logic and the constraints indicate where the requirements for a system to support collaborative design may be further elicited. A summary of the current approach to the analysis and description of teamwork patterns in design collaboration is shown in Figure 3.9.

3.5 Bottom-up flow of information in collaborative design

Following the situation-theoretical framework described in the preceding section, I shall move on to a more detailed exposition of the bottom-up scenario of collaborative design. Starting with the generic items of modelling spaces and modelling acts, I list the situation types in accordance with an action-space matrix. More discussions are then given to each of the situation types derived. Finally, I put forward a scheme of information flow based on the situation types identified, which provides the basis for a discussion of the constraints on design collaboration involving discovering and sharing of common images or metaphors in the next chapter.

Modelling spaces and modelling acts

Several generic kinds of *modelling spaces* and *modelling acts* can be categorised in relation to the scenario description:

Individual Modelling Spaces (IMSs) are the kind of workspaces created and evolved by designers individually for modelling design expressions (e.g., diagrams, drawings, or

Modelling Acts / Modelling Spaces	Representation	Generation	Interpretation	Modeification
IMSs	Individual Object Worlds (IOW)	Local Design Expressions (LDE)	Domain Design Tasks (DDT)	Changes in LDE or in IOW
GMS	Shared integration Schemas (SIS)	Joint Design Expressions (JDE)	Common Design Metaphors (CDM)	Changes inJDE or in SIS

Table 3.1 *A space-act matrix generating eight types of generic design states for describing the situation types in the bottom-up scenario of collaborative design.*

any other graphical/textual constructions) targeted at a particular design aspect or domain of a design project. In collaborative design, there can be many IMSs operated by participants, and by definition, an individual's creating and using his or her IMS may be physically or logically separate from others.

Group Modelling Space (GMS) is the kind of workspaces created and evolved by members of a design team jointly for modelling the integration of design parts as contributed by each team member into larger design wholes. A GMS is initially a *public visual space* for displaying individually-made design expressions; a GMS may be later developed to accommodate new elements and functionality emerging from direct or indirect communication among participants. The emerging design elements and functions (operations) are essential to the realisation of design integration as jointly intended by the team members.

Apart from the modelling spaces, the second dimension of design modelling is about the actions designers perform in the course of developing design schemes. For this, I shall again categorise a number of design actions as *modelling acts*. The kinds of modelling acts listed below are definitely four among many possible others; and I imply no causal relations among the modelling acts.

Representation— the (modelling) acts of forming a design representation scheme with which the actor(s) establishes a *correspondence* between an individual or group modelling space and the ways or mediums of describing or depicting aspects of the artefact yet to be constructed.

Generation—the acts of producing specific (concrete) instances of design expressions (e.g., drawings, graphical models, design specifications, etc.). In short, generation is about the use of a representation scheme by a designer's *intent*.[24]

Interpretation—the acts of assigning, associating, or calibrating the values (or certain meanings) of design expressions generated. The act of interpretation often involves a designer's referring to design knowledge developed and accumulated regarding certain design domains (e.g., building standards, ergonomics, or uses of materials, etc.).

Modification—the acts of making changes in (parts of) the representation schemes established or (parts of) the design expressions generated. The acts of modification normally have the objectives of extending the scope of a representation scheme by introducing new elements or operations or changing the properties and relations of design instances constructed.

A space-act matrix. Recalling my earlier formulation of situation types in terms of the triple of (acts, spaces, states), a space-act matrix can be constructed by combining the two modelling spaces with the four modelling acts which leads to a total of eight design states (Table 3.1). A situation type in collaborative design can therefore be described in terms of performing a kind of modelling act in a kind of modelling space leading to a kind of design state. My next deliberation is to explain what the eight design states prompted by the space-act matrix are about.

Situation types in the bottom-up scenario of collaborative design

Following the matrix shown in Table 3.1, I shall list and explain each of the eight situation types in relation to the bottom-up scenario of collaborative design.

1 [*representation*, IMSs, **Individual Object Worlds**]—Some designer's representation in an individual modelling space leads to the formation of his or her own *Individual Object Worlds*. The term *individual object world* is borrowed from Louis L. Bucciarelli's ethnographic study of engineering design (Bucciarelli 1988), and is used here to denote any object worlds formed by an individual's act of representation performed in an IMS.

Basically, a designer's individual object world consists of a collection of representation elements together with the spatial/functional operations for combining, manipulating, and transforming instances of the design elements as abstracted.

2 *generation*, IMSs, **Local Design Expressions**]—Some individual's generation in an individual modelling space leads to the presence of *Local Design Expressions*.

Instances of local design expressions are design expressions, showing, for example, what spatial forms or particular functions are modelled by the designer when embarking on domain-specific design tasks. Depending on the representation as formed by an individual object world, an instance of local design expression generated in an IMS may have an underlying (conceptual) structure, which defines its parts and part relations. In the case of group working, generations in multiple distributed IMSs may result in local design expressions with possibly highly heterogeneous conceptual structures whose integration may require translations of the participating abstraction schemes.

3 [*representation*, GMS, **Shared Integration Schemas**]—Team members' collective act of representation in a group modelling space leads to the formation of *Shared Integration Schemas*.

This is a situation type in which team members jointly develop shared constructs and methods for the purpose of integrating local design expressions. As said before, local design expressions may have underlying conceptual structures specific to design domains; conceptual structures commonly shared among team members are those constructs and operations that can be called upon to put together parts of local design expressions generated in various modelling spaces. Several such shared conceptual structures found can be further correlated into *integration schemas* which can be employed by group members to handle design integration at a larger scale.

To give examples of group representation in a GMS, consider what is needed for the integration of multiple local design expressions through *aggregation* or *projection* as discussed earlier:

In *aggregation*, instances of local design expressions made in several IMSs are *overlaid* into single expressions in a GMS. Shared integration schemas for facilitating aggregation are mainly concerned with the provision of joint elements and operations that can be employed to (geometrically) interrelate parts of the spatial forms or shapes specified in each of the local expressions.

In *projection*, designers deal with instances of *functions* more than those of shapes. The problem is how multiple functions, as defined by participants in their own individual modelling spaces, can be somehow correlated into new functions with which overall design effects or shapes can be projected in a GMS. This may be explained by that the abstraction of shared integration schemas for projection is mainly through translation and combination of different parts of the functions associated with local design expressions.

Note that aggregation and projection are merely two examples of collective representation in a GMS to acquire shared integration schemas, allowing for combining local design expressions together. There can be many other instances of joint representation in collaborative design. But a general question about group representation in a GMS: Do participants always make shared concepts and methods for design integration explicit? Not to be taken as a definitive rule, the benefit of recording collective design representation or abstraction explicitly is that shared integration schemas arrived at can be reused over and over again during project development. William Lam, the lighting designer for the Cummins Research and Engineering Center, once talked about "repercussions" in multidiciplinary building design (Lam 1977):

Almost every localized decision can be expected to have extensive *repercussions* on the rest of the design. (italicized by the author)

If repercussions are to be kept constantly known among participants making local design decisions, then there is a good reason for the constructs and operations jointly developed by members of the design team to be recorded explicitly; shared integration schemas can be reused by the project team to reveal repercussions whenever changes are made in one domain of local decisions. However, the shared schemas may be disused if new or better ways of putting together local design developments are found.

4 [*generation*, GMS, **Joint Design Expressions**]—Group members' collective generation in a group modelling space leads to the composition of *Joint Design Expressions.*

Given that team members in a group modelling space reach a state of shared integration schemas, joint design expressions can then be generated by applying the shared schemas onto (parts of) local design expressions as put forward by the participating designers. Joint design expressions generated in the bottom-up scenario are in general of two properties:

Pictorial objects. A joint design expression is essentially pictorial (or, at least, diagrammatical), although non-pictorial descriptions or notations may form parts of it. It seems necessary for joint design expressions to appear in pictorial forms on the ground that even if participants do not know much about each other's individual object worlds, graphical communication can still take place among members of the design team. This is an important property as joint design expressions generated in the form of a drawing or a graphic model is more likely to provide a common cognitive basis for evoking participants' interpretations (see below). As seen, all the examples of common images in the case histories (e.g., "vessel," "occulus," "fishbone") can be traced back to some form of pictorial objects.

Graphical structures. Properties of joint design expressions may be further decided by the underlying graphical structures that give rise to the pictorial objects. Broadly speaking, the structures can be of two kinds:

a Single compositions. In the case that common images are generated on the basis of some single unified functions, revealing the overall design consequences of combined local design decisions. The projected waterscape pattern in the Seattle Center Fountain and the fishbone layout in the Cummins Research and Engineering Center are common images of single compositions; no divisible abstract structures underlying the common images that specify parts and part relations can be found. Participants view and manipulate resultant common images in a group modelling space not by parts but as wholes.

b Complex compositions. In the case that common images are generated on the basis of connecting parts of local design expressions geometrically, revealing larger wholes. The coelacanth plan in the Domo Serakanto and the vessel section in the Fitzwilliam College Chapel are common images of complex compositions. That is, there are some abstract structures (as captured in shared integration schemas) specifying parts and relations among parts that constitute the common images. In this case, it is possible for participants to view and get access to parts of a joint design expression in a group modelling space.

5 [*interpretation*, GMS, **Common Design Metaphors**]—Group members' *collective interpretation* of the state of joint design expression in a group modelling space leads to the *emergence* or *discovery* of *Common Design Metaphors*.

This is a situation type in which group members jointly come to a shared recognition of how the common image should be interpreted. In the previous situation type, the participants make efforts to generate joint design expressions in some graphical form, which are not yet evaluated to reveal some design consequences or meanings. Moving onto the current situation, members of the design group reach some agreed interpretations of the joint design expressions in question. Interpretations may be put forward by some individual and then shared by the rest of the design team. There can be two sources of how shared meanings of joint expressions may arise:

a Joint design expressions as a whole *look like* some objects or artefacts familiar to the group members. That is, some sort of morphological or physiognomic resemblance is drawn between the joint design expressions currently developed and some entities in other universe of discourse.
b Joint design expressions suggest some *ways of functioning* recognisable to the team members; for instance, to the mind's eyes of the participants, an environmental system being developed as manifested in the current state of joint design expressions, functions in a similar way to a kinematic or biological system commonly known (or, knowable) to the design group.

It is mainly due to the analogical nature of interpretation as described above that I introduced the term *metaphors* in forming the situation type. Metaphors are commonly used communicative devices when people converse with each other. We should not consider human communication in collaborative design fundamentally different from that in ordinary life situations. Design collaboration does not take place in isolated contexts; even if designers are of different specialisms they come to collaborate on designing buildings with experiences of living in the world that are communicable if not already shared among them. And this is, in my view, the common ground for members of a design team to capture the significance or meaning of their joint design efforts in a metaphorical manner.

However, I should point out that, unlike the metaphors uttered in ordinary conversations, which are, to a large extent, rhetorical devices, design metaphors are likely to emerge *in situ* over perhaps many sessions of collaboration. Firstly, the emergence of common design metaphors is almost materially conditioned by the generation of joint design expressions; secondly, depending on the degree of uniqueness and complexity, it may well take up group members a certain length of time to resolve common interpretations of the joint expressions. Bearing these factors in mind, it seems appropriate to stress the *emergent* nature of common design metaphors in collaborative design.

6 [*interpretation,* IMSs, **Domain Design Tasks**]—A designer's *individual interpretation* of the state of joint or local design expressions in his or her own modelling space gives rise to *Domain Design Tasks.*

This is a situation type in which interpretations of joint or local design expressions take place in individual modelling spaces. Since interpretation of local designs in a local workspace is not directly related to collaboration, I shall focus on interpretation of joint design expressions in individual modelling spaces. An individual's interpretation of joint design expressions is an act of reflecting on the state of local design expressions in the context of an emerging whole. Just to mention two possible occasions:

a Designers verify the part-whole relations between local and joint design expressions that have been assumed prior to design meetings.
b Reflecting on the discrepancies between local and joint design decisions, designers identify alternative part-whole relations yet to be explored after design meetings.

Local acts of interpretation, being different from those of group ones, stand on some domain-specific perspectives from which designers have been working out local design schemes in satisfying certain aspects of the project requirements. The interpretation taking place in local modelling spaces may give rise to information useful to individuals in formulating domain design tasks in which domain-specific design can be further developed.

7 [*modification,* IMSs, **Revised IOWs or LDEs**]—A desiger's *modification* in his or her own individual modelling space leads to changes in the state of individual object worlds or local design expressions.

A situation type where the state of domain-specific design expressions is undergoing modification according to the intents of team members working locally. Or, more fundamentally, the current scope or functionality of an individual object world is deemed by its authors inadequate at some point of time, and subsequently, the domain-specific object world is updated by the authors. These changes can be made by individuals locally without involving other team members in the first instance.

8 [*modification,* GMS, **Revised JDEs or SIS**]—Team members' *joint modification* in a group modelling space leads to changes in the state of joint design expressions or of shared integration schema.

A situation type where the state of joint design expressions is subject to modification as intended by team members working in a group modelling space. If joint design expressions generated are of the complex kind as indicated above, parts of the expressions can be addressed and changed. The acts of making such changes may be performed by some individual but discussed publicly in a group meeting, and participants may agree or disagree any of such changes in the course of the meeting. Similar situations can arise when new integration constructs or operations are deemed desirable to expand or refine the capability of current shared integration schema in use.

In the above, for every single situation type derived from the space-act matrix, I have given explanations accordingly. The space-act matrix, to its best, is a theoretical device that I use to describe some rather abstract aspects of collaborative design activity. An obvious question of doing so is if there exists a reasonable mapping between the theoretical types and the events or activities in the real world. In figuring out the situation types, I have drawn some parallels with some instances from the case histories at various points of discussion. But I do not think a complete matching of the two sets of discussion is any more meaningful then a partial one. The main purpose of the theoretical exposition is to achieve logical consistency in an attempt to describe aspects of design collaboration. On the other hand, the current scope of my case histories is very limited indeed. If the study of case histories is extended to a wider scope, more instances of information sharing or collaborative working relations may be found to illustrate the situation-theoretical types. However, the formulation and exposition of the eight situation types listed above is not the end of the story. In the following chapter, my next task is to chart the flow of information among the situation types. And by identifying the constraints on the flow of information, I shall be able to point out more clearly what the requirements are in developing computer support.

Summary

This chapter starts with a position that an investigation into cooperative design can be founded on analysing naturally created and evolved design expressions taken from historical cases of building design. The aim is to arrive at general conceptual accounts of design collaboration not bound to particular design products or nor to specific strategies of design management. A study of five case histories is presented, revealing a distinctive pattern of collaborative design. Teamwork in architectural design as shown in the case studies seems closely related if certain design concepts, images and metaphors are discovered at some point of time. The concepts or metaphors then serve team members as shared references to emerging wholes, and the shared discovery is in constant interaction with distributed individual interpretations of what parts are and how parts can be interrelated. A bottom-up scenario of design collaboration is proposed as a generic pattern of the collaborative processes. To achieve more rigorous descriptions of the teamwork approach to building design, some basic ideas from the Theory of Situations developed by Barwise and Perry are introduced. We define a situation type in collaborative design as a triple of Modelling Act, Modelling Space and Design State. Eight situation types can be derived following the situation-theoretical framework, which lead to a more detailed exposition of collaborative design involving the emergence of common images or metaphors.

4 More on the Emergence of Common Images and Metaphors

Peter G. Rowe once suggested that various kinds of heuristics are employed to constrain problem spaces in architectural and urban design all in the guise of common types of analogy: "anthropometric analogies, literal analogies, environmental relations, typologies, and formal languages" (Rowe 1987, 80). Rowe defined anthropometric analogies as "mental constructs" describing human's physical occupancy of and movement through a space, and following Geoffrey Broadbent's treatise (Broadbent 1973), literal analogies were of two types, *iconic* and *canonic* analogies. For iconic analogies, Rowe gave the following examples and remarks (Rowe 1987, 82):

> The scope of references for the development of iconic analogies is extremely broad. Objects from the natural world may serve as sources—the shell of a crab for the roof of Le Corbusier's Ronchamp Chapel, hands folded in an attitude of prayer for Frank Lloyd Wright's Unitarian church, sails for Utzon's Sydney Opera House, in a milieu steeped in yachting. Iconic analogies can also include imagery from some scene, painterly conception, or narrative account of real or imagined circumstances.
>
> In all cases the analogy appears useful to designers by virtue of the symbolic or iconographic qualities that they attach to it. It is a physical representation of an intention that, when applied, provides additional structure to a problem.

The case histories in the preceding chapter can be considered as further examples of iconic analogies. Rather than tapping into the analogy to illuminate individual processes of design thinking, our design studies suggest that iconic analogies can also provide additional structures for group communication in collaborative design. To be more specific, the visual and verbal structures serve the design teams as common design metaphors. To recap the previously discussed bottom-up scenario of collaborative design: Successful team working in architectural design seems closely related to if some images, concepts or metaphors are discovered and shared among members of the design team at some points of project development. Starting with the developments of individual parts, members of a design team collaborate on how to bring the distributed design parts together in a satisfactory way. During group communications, some pictorial concepts, images or metaphors are discovered and shared among team members. The common metaphorical referents can be further interpreted by team members individually to make sense of what parts are and how parts may be further modified or articulated with respect to the emerging (architectural) whole. The sharing of the common images or metaphors among group members interacts constantly with distributed individual interpretations of what parts are and how parts should be interrelated.

It has to be said that there can be more than one way of describing and interpreting patterns or approaches to collaborative design even based on the same set of case histories. The purpose of going through these case studies here is mainly to do with the elicitation of (informal) requirements for computing support in collaborative architectural design. To achieve this, I have developed a situation-theoretical exposition of the bottom-up scenario in the preceding chapter. According the situation theorists, the analysis should lead to a description of the logic of activity being studied. To follow the same line of enquiry, I shall in this chapter spell out the constraints on the flow of information among the situation types identified earlier in the exposition. The constraints elicited are then viewed as pointers to the issues and requirements for developing computer support in collaborative architectural design.

4.1 Constraints on collaboration in the bottom-up scenario

Assuming a design team consisting of three designers working on separate aspects of a building project, a diagram can be drawn by relating the situation types elicited from the bottom-up scenario (Figure 4.1). The diagram is to be read from bottom up, and the pointing direction of the arrows in the diagram indicates also the flow of time starting from the bottom. The diagram is drawn not to illustrate an entire collaborative design process but to represent perhaps a typical session of collaborative working. Starting with designers' representations of individual object worlds and generating local design expressions in their own individual modelling spaces, the diagram shows a session of generating and modifying joint design expressions in a group modelling space.

In the following four sub-sections, I shall continue to examine the flows of information among the situation types as shown in the above diagram. An indicative question is raised regarding the condition or involvement required for information to flow from one situation type to another as suggested in the diagram. Through a closer look into these questions, a set of generic constraints on the flow of information in the bottom-up scenario of collaborative design is described in more concise terms.

1 [*generation*, IMSs, **Local Design Expressions**] → [*representation*, GMS, **Shared Integration Schemas**]

From distributed local design expressions ($\boldsymbol{LDE}_a$, $\boldsymbol{LDE}_b$, ...) to collective design representation in shared integration schemas (***SIS***)—i.e., information flow from the situation (type) where multiple local design expressions generated by design team members in individual modelling spaces to the situation where shared integration schemas for connecting local design expressions are jointly specified by the team members in a group modelling space.

Question: *What is involved for members of a design team to arrive at shared integration schemas following that they have produced local design expressions in distributed individual modelling spaces?*

***Figure 4.1** The flow of information among the situation types derived from the bottom-up scenario of collaborative design. (Note that the number of designers shown on this diagram is only an assumption; in theory, the number can be scaled up to any group size.)*

Recalling the case histories seen in the previous chapter, let us consider the following conditions that participating designers might have to work with in order to arrive at the integration schemas to be applied to the local designs in a group modelling space:

a A *communal visual space* is required for displaying all participants' local design expressions such that they are able to perceive and discuss the potential interconnections among parts of the expressions jointly presented. The communal visual space would admit instances of local design without the appearances of the expressions being altered or translated.

b A set of *spatial constructs* and *spatial operations* are construed and specified jointly by participants such that domain-specific constructs and operations that are employed to produce local design expressions can be merged to form schemas for integrating graphical forms or geometric shapes contributed from distributed local workspaces. A *common language for spatial modelling* is required, which provides the basic building blocks for group members to arrive at integrative spatial constructs and operations that can be reused whenever there is a need to (re)combine and transform multiple local design expressions into single joint schemes.

c A set of *projective* or *generative* rules is construed and specified in group modelling space that is able to take local design expressions as input and project or generate joint design expressions as output. The joint abstraction leads to schemas for integrating the functions or performances of design parts proposed by participating designers. A *common knowledge base* is required, which provides the design team with tools for functional modelling that can be called upon to generate depictions or descriptions of overall system behaviours based on local parameters.

Clearly, the above three conditions point to the need of holding *design meetings*, in which local design schemes brought by different individuals in distributed object worlds are gathered together in a communal visual space. Assuming that during the design meetings the participants come to recognise some shared goals of integrating individual contributions, a group modelling space is required to provide shared platforms for performing joint spatial and/or functional modelling. Regarding this, I shall denote the constraint on the information flow from local design expressions to shared integration schemas in the following expression. A summary of all the terms and notations used to denote the constraints is provided in the Glossary.

Constraint 1: ([***LDE***$_a$, ***LDE***$_b$, ...] $\rightarrow$ ***SIS***) $\Rightarrow$ Communal Visual Space & Group Modelling Space containing common language for joint spatial modelling, common knowledge base for joint functional modelling

2 [*generation*, GMS, **Joint Design Expressions**] $\rightarrow$ [*interpretation*, GMS, **Common Design Metaphors**]

From joint design expressions (***JDE***) to the discovery of common design metaphors (***CDM***)—i.e., information flow from the situation where joint design expressions are constructed in a group modelling space to the situation where common design images or metaphors are discovered and shared by group members.

Question: *Given that joint design expressions are constructed by applying shared integration schemas onto local design expressions gathered in a group modelling space, what is involved for members of the design team to arrive at shared interpretations of their integration?*

The case studies show that the sharing of design metaphor is a continuously interactive process. When a concept of wholeness or uniqueness is reached at some stage, it, in turn, provides the team members with a cognitive basis to articulate the existing designs of parts (taking the example of the Domo Serakanto project, designers examining the parts of 'gill,' 'spine,' 'horn,' 'cilium,' 'teeth,' 'antennae' and 'scales' within the 'coelacanth' as a whole). And it is possible that some members' articulations may give rise to new parts to be developed, which can consequently lead to a new state of the wholeness. Referring to the designers' recollections in the case histories, I should point out that not for every instance of joint design expression generated there is always a discovery of common design metaphors. If the group processes do occur in the end, it may take the design team many sessions of co-design to arrive at such a discovery. But what are the general conditions for common images or metaphors to emerge even occasionally? Given the evidence from the case studies, we can perhaps infer the following:

- A cognitive agent's *analogical reasoning* is required to first recognise the resemblance or association that exists between the current state of joint design expressions and some concepts or images that may belong to some other worlds of activity.
- A cognitive agent's *explaining* ability is required to communicate the analogy found in a way that is understandable to other members of the design team such that the team as a whole comes to agree such an analogy programmes to report on in separate volumes. I therefore limit my current discussion to merely pointing out the requirement for cognitive agents who are able to perform analogical reasoning and/or association.

It should be noted that a cognitive agent could be either a human designer or a machine agent as in artificial intelligence. Analogical reasoning and explanation as cognitive processes are established subject areas in recent developments of cognitive psychology, computational intelligence and computer-aided design (see, for example, Barnden and Holyoak 1994; Coyne 1995; Vosniadon and Ortony 1989). A thorough treatment of analogical reasoning and associative thinking in the context of collaborative design is beyond the current scope of this book, which deserves a dedicated volume to survey this extremely rich research field. Seen from our current study, the second constraint on design collaboration can be described in terms of the capabilities of performing analogical mapping or reasoning whenever there are 'cognitive leaps' from joint design expressions to common design metaphors. I shall discuss more about the supporting issues in relation to this constraint in the next section.

Constraint 2: (*JDE* → *CDM*) ⇒ Analogical mapping, reasoning and explanations

3 [*interpretation*, GMS, **Common Design Metaphors**] → [*interpretation*, IMSs, **Domain Design Tasks**]
From common design metaphors (**CDM**) emerged in a group modelling space to *distributed domain design tasks* (DDT_i)—i.e., information flow from the situation where common images are discovered on the basis of joint design expressions generated in group modelling space to the situation where participants identify domain-oriented design tasks in responses to the implications of common images for local design developments.

Question: *Under what circumstances do participants identifying domain design tasks with regard to the current state of joint design expressions and its interpretation become problematic?*

The discovery or emergence of common images and team members sharing the interpretations can be potentially a breakthrough to the design team in project development. However, the significance of the discovery cannot be determined sharply by someone having a moment of eureka; common images and metaphors can be significant if and only if they are deemed by the design team as *prompts* for further local design actions, and the team members will have to play out the implications of the images from domain-specific perspectives. It is until group members succeed in identifying domain-oriented tasks for exploring design developments further that the sharing of common images among team members comes to its fullest sense. Consider the following conditions in which design participants may find the situation problematic even if some instances of common images or metaphors have been proposed during a design meeting.

- Participants cannot agree upon how and where newly generated design parts or relations among existing and emerging parts (due to design integration) should be taken care of.
- Some group members working in local domains find themselves incapable of handling the transformation, extension or translation required to further develop their local designs in individual modelling spaces in order to sustain the sharing of common images.

The conditions above point to the possibility that the emerging parts in joint design expressions are beyond the current ranges of existent individual object worlds. To make the group working continue to flow, the design team may decide to abort the current attempts and start afresh another route of integration; or, the discovery still looks so promising to the team that a new member is brought into the team so as to expand their joint expertise in tackling the newly found areas of project development. In perhaps a less drastic measure, some team member may be delegated to expand his or her individual object world in order to take on the emerging parts as an added design responsibility. All these considerations seem to suggest that participants' identification of domain design tasks in relation to the discovery of common images is constrained by if the design team as a whole can provide adequate ranges of individual object worlds for handling emerging parts and relations.

Constraint 3: (**CDM** → [DDT_a, DDT_b, ...]) ⇒ Adequacy in the ranges of individual object worlds

4 [*modification*, IMSs, **Changes in Local Design Expressions**] → [*modification*, GMS, **Changes in Joint Design Expressions**]

From *design changes in local design expressions* ($\Delta\sum LDE_i$) to *design changes in joint design expressions* (ΔJDE)—i.e., information flow from the situation where some changes are made intentionally or consequentially in local design expressions by one or several participants working in perhaps various domains to the situation where joint design expressions are modified either consequentially or intentionally.

Question: *On what occasions where changes made in parts of local design expressions cannot be realised in relation to the changes made in joint design expressions, and vice versa?*

Given that adequate share integration schemas are operational in a group modelling space, participants making changes in their local designs, for whatever reasons, should lead to *consequential* changes in the joint design expressions as they are generated by applying the schemas onto local design expressions in the first place. By reapplying the schemas onto changed local designs, participants are expected to see changes to occur in joint designs. Design changes can be initiated from the other way around; that is, parts of joint design expressions are intentionally modified by some group members for whatever reasons.[25] The intentional changes in joint design expressions can have consequences on local designs if the designers involved are aware of the changes.

However, it is likely that changes may not be followed through in concerted steps either from locals to the joint or from the joint to the locals under various circumstances in which the flow of information breaks down. If shared integration schemas are deployed by group members as common working protocol for design integration and distribution, changes intended by some individuals may have important implications for the works pursued by others; that is, the *repercussion* effect. For design participants to actually *implement* their intended changes in parts of local or joint designs, their intentions have to be publicized to other participants and request for holding an *exploratory* or *tentative* design integration/distribution. By examining the *changing* state of joint/local designs, participants evaluate the potential consequences on their domain design works; they may subsequently support or reject the changes proposed by their colleagues. Consider the following cases of rejection:

- *Backtracking.* The designer who proposed changes has to drop the intended changes because some members cannot accept the outcomes or the implications of the proposed changes from their domain-specific design perspectives.
- *Competing.* The proposed design changes are deemed doubtful by some members but the proposal invites or provokes other members' design thinking, and they may subsequently propose alternative design changes that may in effect compete with the earlier ones.

If proposed design changes are judged acceptable or desirable by all members of the design team, events to follow to complete design changes in all related aspects may include:

- *Coordinating.* Participants are in favour of the implications from an exploratory integration and respond with making necessary changes in local design expressions in a coordinated manner.
- *Confirming.* After participants take part in coordinated design changes, exploratory or tentative design changes are declared as *confirmed* changes by participants involved in the current session of coordinated design changes. The design team can then move onto other individual or joint design tasks.

To summarise the discussion above regarding making design changes, the constraint on the flow of information among distributed individual modelling spaces and a group modelling space is denoted as follows:

Constraint 4: $(\Delta\sum \mathbf{LDE}_i \rightarrow \Delta \mathbf{JDE}) \Rightarrow$ Joint evaluations of tentative integration or distribution of proposed design changes

In the above, we have looked at four generic constraints according to the situation-theoretical analysis. In theory, there can be more flows of information involving other types of situations elicited from the bottom-up scenario of collaborative design. However, an exhaustive enumeration of all possible flows of information among all the situation types will be extremely tedious, and some of them may not reveal critically about the constraints on team-working processes. Methodologically speaking, collaborative design as a human endeavour and its logic can be studied in relation to perhaps many other theoretical frameworks. I do not intend to imply that an application of the Situation Theory concludes in any way a practical model of collaborative design processes. Instead, the situation-theoretical framework[26] has been used to elaborate on the constraints on design collaboration, revealing an internal logic of collaborative design. The current study shows that designers working in a way similar to the bottom-up scenario may not know much about or, indeed, are not concerned with each other's domain-specific design developments, but they can co-work as a team if continual communications and interactions between distributed local designs and resultant integrated designs can be sustained through a project's lifetime. If it does occur, the discovering and sharing of common images and metaphors can provide the design team with useful or even powerful strategies in architectural thinking specific to the project.

4.2 Issues in supporting emergence of common metaphors

In Chapter 2, we have seen examples of early experimental systems that were developed to support collaborative drawing and design in various ways. Clearly, some of the collaborative drawing tools may be deployed to serve some of the group design processes discussed in this chapter. For example, designers may find collaborative drawing tools like *Commune* or *GroupSketch* useful when engaging in graphical conversations to resolve differences or conflicts in making design changes. However, these tools are limited if the aim is to support more sophisticated design operations such as exercising *individual object worlds*. As discussed earlier, a designer works with his or her own individual object world when creating local design expressions, which in turn are the basis for generating joint design expressions via shared integration schemas. In this regard, the constructs and methods of creating design expressions are just as important as the outcomes of design drawings or models. There are other aspects of group design communication that remains to be addressed, which may lead to the development of new system features in computer-supported collaborative design. In this final section, I shall move on to discussing a set of five issues on developing computer support for collaborative architectural design following the constraints studied in the previous sections.

Support for meeting of individually developed design parts

The meeting of parts implies the use of a group working space in which members of a design team collaborate to put their individually designed parts together. In a traditional manual-based setting, this may simply take place around a table where pieces of drawings or models are brought together. If we think of a computer-based approach, i.e., team members working on individual parts with computers, a computer-supported group working space will be a desirable facility. Assuming that computer networks and proper software are in places, a computer-based group working space can be created in at least two ways: (1) group members are collocated and have face-to-face meetings augmented by a network of workstations with or without a common projection screen in the meeting room, or (2) group members are geographically dispersed and have *virtual meetings* over a computer network which enables participants to share drawings and other kinds of data in real time.

What is required is basically a digital *communal visual space*, as mentioned earlier, in which members of a design team can jointly present their (up to date) local design expressions so that potential spatial and/or functional connections among individual work can be envisaged by whoever participates in the meeting. In this regard, typical operations to be supported include:

Networking. In developing design solutions to satisfy domain design requirements, participants create and evolve their own modelling spaces; these individual modelling spaces can be distributed over several remote working sites, and participants may have different preferences of when to work. A *common visual space* should be operating synchronously or asynchronously among all participants' working sites, providing designers with immediate access to an essentially shared graphical view of what design solutions have been explored in related domains.

Group meeting scheduling. At any one time, one or more participants may request to hold a face-to-face or virtual design meeting to keep up to date the interrelations among local design developments. Requests for project meetings may indicate a possible time and intended agendas of the proposed meeting. Messages of *call for meeting* are generated and sent to possibly remote sites, asking for participants' replies; when replies are gathered and processed, a *project-based meeting scheduler* should broadcast a meeting schedule to all participants involved.

Graphical structure filtering. Since local design expressions are constructed by participants working with their own individual object worlds specific to their domains of design expertise or space-conception processes, the conceptual structures underlying these locally constructed graphical expressions can be *heterogeneous*. When imported into the communal visual space for design meeting purpose, instances of local design expression need to be converted to *filtered images* in order that the domain-specific structures are left out while preserving the surface features of each expression contributed. The graphics filtering facility should be provided in connection with the communal visual space.

(b) Support for joint authoring of shared integration schemas

The communal visual space is a general-purposed, multi-user graphical workspace for gathering local design expressions. A (computer-based) group modelling space should be made available at a time when group members gather together to generate integrated design expressions on the basis of shared integration schemas. The question is how shared integration schemas become available for the design team to visualise the outcome of putting local design parts together. Since designers operate with their own individual object worlds, it is not possible for a collaborative design modelling environment to provide such integration schemas in the first place. In practice, shared integration schemas will have to be co-produced by participating designers in a way not dissimilar to the generating of individual object worlds. To this aspect of group working, what a collaborative design environment can offer is a *common language* for co-authoring shared design integration schemas that can be used repeatedly. A *common language* can be considered as an abstract design modelling language, providing a general representation framework that designers can learn to work with. To facilitate group authoring of shared integration schemas, a common design modelling language can be developed to provide a set of basic computational constructs (geometrical and/or algebraic) to achieve the following:

- Specific constructs can be co-authored to enable group modelling of newly identified design elements or functionalities in the course of exploring how locally produced design parts may be combined satisfactorily.
- Specific production rules as transformation or projection procedures can be co-specified and applied to the interconnecting of existent domain-specific constructs used locally.

(c) Support for the discovering of common images and metaphors

In general terms, common design images or metaphors emerging from teamwork in design are pictorial and/or linguistic expressions of some holistic significance or meaning derived from the meeting of design parts by members of the design team. Given drawings or other modes of visual representations that depict combinations of design parts, team members may or may not reach a shared recognition of a whole out of a sum of the individual parts. In cases where a common recognition of wholeness, order, or unity is obtained at some point of project development we can say that the group members have reached a status of sharing common design images or metaphors.

The kind of metaphors appeared in collaborative design by discovery is somewhat different from the concepts and uses of metaphors studied by linguists (see Lakoff and Johnson 1980 and Mori and Nakagawa 1991, for instance). The linguists tend to view metaphors as largely existent linguistic entities that are used by people who are having conversations. As shown in the case histories, the metaphors shared among designers in team working are essentially picture- or image-oriented. Though there can always be some linguistic terms associated with a metaphor (e.g., "vessel," "wall," etc), the usefulness of shared metaphors in communicating about building design is distinctively and specifically imagistic.

It can also be pointed out that, giving a project brief to different design teams, even if they reach the same metaphor, say, "coelacanth," it is likely that they actually produce distinctly different designs because the very parts that constitute the fish are differently designed and the joints developed to bring parts together may be varied. Furthermore, we have seen in the Seattle Center Fountain project in which the common image (i.e., the projected waterscape over the score and diagram) developed is basically a graphical pattern without a specific name.

To make another point, unlike metaphors appearing in ordinary conversations, which are, by and large, existing rhetorical devices, design metaphors seen in most cases are discovered or developed in situ after certain sessions of group working.[27] The discovering of common design metaphors is conditioned by some member(s) seeing and extracting features from an aggregation of design parts. Depending on the level of complexity presented and abstraction required, it may well take group members a certain period of time to reach a commonly understood and agreed interpretation of the emerging whole; and to communicate effectively the finding, the abstraction has to be captured by some metaphorical and graphical devices.

In all the case studies, no clear evidence shows that common design metaphors are represented in any explicit ways, but are seemingly kept in the team members' heads. What computing tools can be supportive of the discovering of common images/metaphors in collaborative design? My current observation suggests that features extraction and interpretations (or, mapping) can be relevant. In general, features extraction leading to the discovery of common design metaphors in collaborative design can take place in two possible ways:

- The combined parts as a whole looks like some natural objects or otherwise designed artefacts to (some) members of the design team; for instance, some kind of morphological or physiognomic similarity or resemblance is recognised between the collection of parts currently aggregated and the other things referred to.
- The assembled parts as a whole suggest certain ways of working (or, functioning) recognisable to the team members; for instance, to participants' mind, the system being developed, as manifested in the current attempt of putting parts together works in a similar way to, for example, an existing kinematic or biological system that are commonly known to the design team.

Given the above creative and emergent nature of metaphor-finding, it is difficult to suggest what computational tools may be decisively useful to the analogical mapping processes. Features extraction techniques developed in pattern recognition may be relevant in this aspect, but any mapping between syntactic and semantic levels will impose a certain degree of *prescriptiveness* when building the systems which can be undesirable from an architectural design point of view. Systems that function only at a syntactic level, however, will not be immediately useful from the viewpoint of design supporting. This seems to suggest that this critical stage of team working in design remains largely in the realm of human designers' cognitive capabilities of drawing analogies and imagination.

(d) Support for communicating common images or metaphors discovered

The immediate issue next to the discovery of common images from meetings of design parts is the communication of the findings among team members. An image or metaphor discovered will become common if and only if a majority of team members are able to appreciate and agree with that particular interpretation. As pointed out earlier, the media used to convey the finding will be essentially visual or graphical. 2D images/graphic patterns or 3D object/spatial models are likely to be used for the purpose of explaining where the group work is leading to as a whole.

In facilitating group communication in explaining emergence of common design images or metaphors in collaborative building design, computer-based resources and tools can have a greater role to play. What would be useful to the design teams is basically a computer-based database allowing participants to quickly retrieve instances of visual references that may correspond to the images emerged. However, given the emergent and non-deterministic nature of discovering common metaphors in collaborative design, the database will have to be substantially large to be useful and it should enable users' search for references. In principle, the databases should be as large as a reflection of the world that we live in; this implies that nearly every aspect of human experiences and knowledge are covered by the systems. The scope of the systems required is clearly beyond the current ability of any single enterprise.

A possible candidate as a technological response to the requirement above is the Internet and the World Wide Web. The continuous robust growth of the Internet and WWW is virtually participated by any individual around the globe who can get access to the global network. In the next section, I shall have more detailed discussion of the potentials as well as pitfalls of this global digital online resources in relation to the processes of collaborative design by discovery.

(e) Support for coordinating design changes and distributed local design developments

If the consequences induced by a proposed design change are considered desirable and manageable, one designer's making changes in one place may cause further changes to be coordinated by designers working in related domains. Depending on how shared integration schemas are co-specified by team members, a *housekeeping*-like mechanism can be deployed to assist design consultation among participants for detecting and resolving potential conflicts as manifested in the changing states of common images. More specifically, a computer-based bookkeeping agent may behave in the following ways to support coordination among geographically dispersed co-designers:

- Detecting state changes in joint design expressions arising from local changes proposed by any participating parties
- Knowing and locating whomever needs to be consulted whenever a state change in joint design expressions is detected
- Delivering all responses from members involved in the consultation to the designer(s) making the proposed changes.
- Sending reports to the right participants requesting confirmations or other modes of coordination.

4.3 The World Wide Web and collaborative design by discovery

I propose in the above that an online digital repository containing more than enough visual references can be potentially useful to designers when communicating and explaining discovered common images or metaphors. There are two basic issues here regarding such a digital information system: (1) size or boundary of the system, and (2) information search and retrieval.

The amount of visual references potentially relevant to innovative collaborative design teams ought to be enormous because designers may draw on any images as analogies that are interesting to them; that may range from records of artefacts attributed to some of the oldest human civilisations to pictures captured by the latest scientific discovery. Therefore, it is inconceivable for such databases to be purposely built in support of collaborative building design as the boundaries of such systems have to be kept open as long as the bodies of human experiences and knowledge continue to grow. Nevertheless, we do see designers in various fields building *libraries of slides* in addition to libraries of books and design journals. And it is not unusual to see designers talking to each other by referring to images on books or magazines as a way of getting design ideas across. Moving onto the era of digital studios, I expect the same purpose of communication will be served by digital means, i.e., the building and using of digital image banks.

The enterprise of making of electronic encyclopaedia of images as a kind of online resource is potentially relevant to design practice in many fields. The archive project "Corbis" launched sometime ago through a private initiative intended to deliver a digital image library that will contain millions of digital images covering various realms of human endeavour (Rapaport 1996). However, I shall contend that a centralised data bank maintained by a single private enterprise or institute will inevitably be imposed with an artificial boundary biased toward commercial incentives or market constraints. An alternative of having an online resource of this nature, which is free from time, market and other constraints, is to have distributed developments of the database which can be participated publicly by as many individuals, groups, and institutes as possible. In fact, a public infrastructure together with other technologies have already been made available for on-going distributed accumulating of digital information resources at a global scale—the Internet and the World Wide Web[28] (Web, in short).

Apart from the continuous growth of the Internet[29], the ability of the current Web technologies to handle various kinds of visual information is the most obvious reason that I consider a potential connection between Web-based resources and collaborative design by discovery. People world-wide can now deposit, search and retrieve digital visual information in terms of graphic images, video clips, animation, 3-dimensional virtual reality models, and combinations of some of those. All an end-user needs is an access to the Internet and a hypermedia-enabled Web browser.

I also take the view that the various general-purposed "search engines" currently available on the Internet (such as *AltaVista*[30], *Excite*[31], *Lycos*[32], and *Yahoo*[33]) is another facility provided in the Web that is potentially supportive of collaborative design by discovery. These Web sites of search engines are themselves pointers to other sites that may contain information of interest. Opening up the search sites, any users can type in a key word or phrase and the Web sites will return the users with the hyperlinks to all other sites that contain the keyword or key-phrase sent. As a matter of curiosity, I have tried "coelacanth" on one of the search sites, and the image below is one of the references retrieved from the Web site maintained by The American Museum of Natural History (Figure 4.2).

To me, this illustrates how the current state of the Web may actually facilitate designers' searches, retrievals, and presentations of visual references for communicating their discoveries of common images in collaborative design. In many ways, researchers and developers worldwide are just beginning to tap into the potentials of combining the Internet as a globally distributed information hub and the Web as unifying interfaces for editing and viewing documents in diversified digital media (Kim 1997). In the years to come, we can expect that online referencing will become just as indispensable as physical books and journals in most design practices.

The usability of the Internet and the Web to facilitate collaborative design by discovery of common images or metaphors will not be known unless the technologies and resources are evolved to a state that the reliability and quality of the visual references are of a widely acceptable standard. There is also the problem of search. Currently, images can only be searched in terms of how they are named or indexed. That is to say, we cannot simply sketch, draw, or scan any images and feed them into a search engine and expect it's returning images that bear some resemblance or relevance to the ones we are interested. The technologies for image-based information retrieval or visual search still have some way to go to reach wider applications (Chang and others 1997; Gupta, Santini, and Jain 1997), which are inherently rich and complex to even define the functionality of image recognition and retrieval in standardised terms. Unlike words and other linguistic entities, pictures or images are much fuzzier and subject to many different interpretations depending on who is watching. If any significant progresses are to be made in this area, it will probably start with some domain-specific subject matters with more restricted definitions of how pictures are formed and interpreted.

Given the open and participatory nature of *growing* the Internet and the Web, we may still feel promising that the Web-based information resources can be a useful online facility to aid group communication in architectural design even in a limited manner. Drawing on the ever growing amount of visual information resources distributed among a vast number of Web servers connected by the Internet, it can in theory serve collaborative designers in presenting and explaining instances of common images and metaphors with perhaps a higher degree of effectiveness and immediacy. However, the

Figure 4.2 An image of coelacanth. (Copyright 1996 Source: The American Museum of Natural History, from http://www.amnh.org/Exhibition/Expedition/Treasures/Coelacanth/coelacan.html).

speediness, ease of access, economy, and convenience offered by the Web may foster work habits that deprive designers of critical minds and first-hand bodily experiences and become withdrawn from the real world as expounded by Christine M. Boyer (Boyer 1996). Perhaps we should constantly remind ourselves that, just like conventional books and magazines, 2D or 3D digital images shown on the computer screens are no replacements for getting direct contact with real things in the real world, which will always remain essential in communicating design discoveries.

4.4 Intranets and Web technologies for Building TeamCAD

At the beginning of the book, I state that practice-pulled reflective design studies can also make contributions to the research and development of teamwork-enabling technologies. Taken as examples of group practice in design, a number of case histories of team design are studied to gather what designers could recall about their experiences of designing with others. In search for similarities across the case studies, I present an explanatory theory of collaborative design by discovery. It should be stated again that collaborative design by discovery is by no means the most typical way of team working in the field of architecture, nor is my current study intended to lead directly to any specific implementations of a computer-supported collaborative design environment. Nevertheless, following an analysis of the constraints on the bottom-up scenario of design collaboration based on the situation-theoretical framework, I was able to point out at least some of the issues of supporting collaborative design if a development of computer-based facilities is to be considered.

Unlike many other fields of design, architectural design has to respond to particular contexts: the site, the time, the clients and the technologies for construction. This fundamental demand for specificity and spontaneity in architectural design contributes to a high degree of *dynamics* in a parts and wholes relation that design teams must deal with. Members of a design team produce different parts, and parts are put together to form greater parts as the project develops. The integration of design parts is governed by if some sort of *architectural wholeness* can be found and shared among members of the design team. I have characterised the process as the emergence or discovery of common images and metaphors.

I have also considered that the World Wide Web can serve as online open repositories in support of collaborative design by providing team members with visual references in communicating part-whole relations. Given its being open to continuous participatory development, the WWW has evolved into a significant and popular forum for developing information sharing and retrieval technologies. In fact, a family of Web-oriented technologies has already emerged which are currently used in the IT industry to build the applications of *Intranets*. An intranet can be considered as a *mini* Internet with a clearly defined network boundary set by a company or an organisation on the basis of the Internet. The uses of an intranet are internal to the ownership of the network, which is neither visible nor accessible to unregistered or unauthorised users. However, an intranet can be constructed with the same set of computing communications protocols as the Internet such as TCP/IP (Transport Control Protocol/Internet Protocol) and HTTP (HyperText Transfer Protocol). What is required further in order to define an intranet's boundary is a network management regime that imposes certain schemes of digital registration and authentication. Given the feasibility of applying the infrastructure technologies in building a collaborative design environment, we may think of an intranet to serve as the backbone for a project-wide network communications with which members of a design team can perform design tasks jointly in design abstraction, integration, articulation, and coordination.

To conclude my current study of teamwork in architectural design, I would like to use the diagram shown in Figure 4.3 to reiterate what I consider the key points about, firstly, how collaborative design as a natural activity may be explained, and secondly, what mechanisms and facilities of computing communications may be useful to support design collaboration.

The diagram above should not be taken as a blueprint for building a system architecture of any kind. It rather indicates a set of general requirements to be considered when developing a team-oriented computer-aided design environment.

- *Synchronous or asynchronous team working.* Designers may collaborate with one another in both synchronous and asynchronous modes. Synchronous communication may be not necessary when team members are engaged in domain design tasks at earlier stages of a project. However synchronous group processes can play an important role at later stages for the design team to reach a common understanding or projection of how individual design contributions may converge into a satisfactory and coherent architectural whole.
- *Flow of information.* The flow of information is primarily *bottom-up* oriented; that is, from local design decisions in multiple distributed individual modelling spaces to integrated designs in a single shared group modelling space. Design collaboration is mainly driven by participants' development of domain-specific design parts. When the participants find a way of putting parts together in a meaningful way, they may thus get better ideas of how the parts can be further developed. Making design

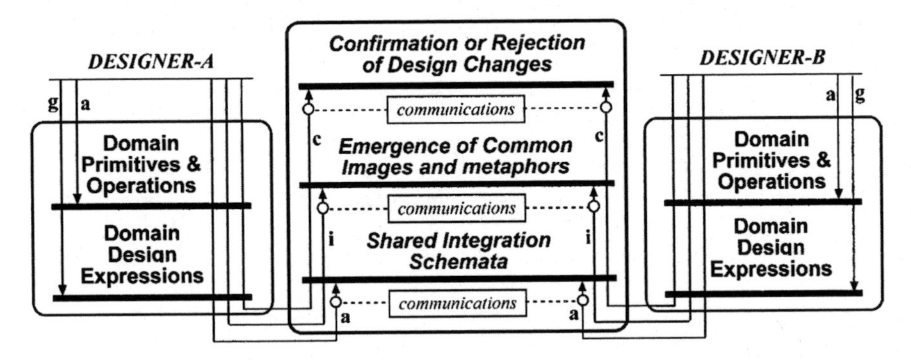

Figure 4.3 A schematic diagram to summarise the bottom-up scenario of design collaboration. The number of designers indicated here is arbitrary. (The scaling of 2 to N designers in a design team can be envisaged by viewing this diagram as a section of a cylindrical structure. The diagram also suggests that an individual's workspace is a combination of his or her own IMS and a GMS.)

changes is also coordinated at local levels with reference to the consequences of changes as reflected in design integration.

- *Individual and group modelling spaces.* The provision of both individual and group modelling spaces is essential to sustain collaborative design. The design team may consist of markedly different individuals who may work in different domains of a building project, employing specialist object worlds in developing designs. The team's group modelling space is capable of supporting (1) a communal visual space for visualising the collection of design parts without structural integration in the first place, (2) group authoring of shared schemas for spatial and/or functional integration that govern the production of joint design expressions, (3) exploratory or tentative design integration by showing potential consequences of proposed local design changes which may or may not be implemented in the end.
- *Internet and/or intranet.* A combination of multiple distributed individual modelling spaces and a shared group modelling space can be implemented as a project-wide intranet that can facilitate distributed asynchronous individual design working as well as remote synchronous virtual design meeting sessions. Connections with the Internet can be desirable for the design team to get access to online digital visual references in support of discovering and explaining common images or metaphors among members of the design team.

Summary

Following the situation types of the bottom-up scenario listed in the preceding chapter, we look into the constraints on collaborative design. Constraints are identified through analysing the conditions for information to flow among the different situation types. The current analysis suggests four clusters of constraints on design collaboration: (1) the provision of a Communal Visual Space in group modelling space supporting common visual languages for joint spatial modelling and common knowledge bases for joint functional modelling; (2) the exercising of analogical mapping/reasoning and explanations; (3) the adequacy in the ranges of individual object worlds employed; and (4) the joint evaluation of proposed integration or distribution of design changes. In accordance with the constraints examined, we discuss some of the issues concerning the development of computer-based facilities to support aspects of group design communication.

Among others, the access to online visual resources is considered useful to facilitate the discovery and explanation of common design metaphors. This points to the potential of the Internet and World Wide Web as the online repository that is rich and dynamic enough to reflect the world we live by. Some researchers have explored how to search and retrieve images from large digital libraries of visual references, which may be integrated into future Internet-based search engines. Comparing with the early CSCW systems, the recent rapid developments in *intranet* applications are steps forward to delivering computing communications in building TeamCAD environments that can supportive of a project team's collaborative design by discovery.

5 Flexible Generic Frameworks and Collaborative Design

To continue our study of collaborative design in the field of architecture, this chapter presents an analysis of a different scenario of team working in design. Like before, I shall present a number of building design case histories that show how representations of building design and modelling can be interrelated with communication among members of the design teams. We shall see how members of a design team collaborate to achieve integrated design on the basis of sharing and substantiating common *generic flexible structures* with distributed domain design developments. Examples of generic flexible structures drawn from the history of architectural design are introduced, showing that teamwork can be organised by participants sharing some kind of generic frameworks or objects. In articulating the features of collaboration, *a top-down* scenario of collaborative design is presented. The scenario is then treated with a situation-theoretical analysis in some detail. A set of constraints on collaboration are elicited, pointing to a different logic of collaborative design: the need of maintaining a dual correspondence between the group modelling of common flexible generic structures and the development of domain-oriented designs distributed among multiple individual modelling spaces. In the final section, mainly because of the extensive use of the term *structures* in this chapter, I draw some connection with the Dutch architectural Structuralism—an influential school of architectural thought first introduced by the Dutch architect Aldo van Eyck in the late 1950s.

In an interview by Abbey Suckle, Richard Rogers recalled an important stage of design development in the Centre Pompidou (or, Plateau Beaubourg) project in which the leading design idea of "an architecture of possibilities" was to meet up with the Parisian codes of fireproofing in building structures. It turned out, as Richard Rogers reflected, that the design team did not succeed in finding an elegant solution to the problems of how to make the steel structure incombustible. Instead, to satisfy the required fireproof rating, the design was developed in an ad hoc way ranging from the reduced building height to the protection of the steel truss beams with fibrous wrapping and stainless steel encasing. Regarding this well-known case of a building project, Peter Rowe remarked, "Indeed, the visual effect of almost every element in the building had to be thoroughly considered from this perspective." (Rowe 1987, 34)

As the technological palette for building design is being developed even more rapidly nowadays, the story above illustrates no longer an exceptionally rare situation in contemporary design practice. A team is formed by congregating designers of different disciplines, who work in separate design domains and employ specialist tools and knowledge demanded by the clients or users. However, it is also essential for the design team to arrive at some *architectural wholes* that address all requirements adequately and beyond. Arguably, architecture is one of the design fields that values creative synthesis more than any other endeavour. The problem for every design team is how to achieve successful synthesis of built form on the basis of multidisciplinary design inputs with a high degree of efficiency. The Centre Pompidou project also demonstrated that it is rather unpredictable in innovative building projects what design constraints will prevail that eventually becomes the major source of forces to which the character of the building is shaped to respond. Multidisciplinary or interdisciplinary collaboration is therefore crucial if a building design is not to be, in Richard Rogers' words, "divorced from its legal, technical, political, and economic context," and architectural teamwork should be managed to allow constant flow of the team members' spontaneous insights so that "constraints may be absorbed and whenever possible inverted into positive elements."

Designers of the built environment normally produce representations of objects or spaces yet to be built in the real world. In the preceding chapters, we see uses of representations of buildings or other built environments by designers not only to delineate the edifice to be built but also to mediate collaborative working among members of a design team. In the following section, we shall look at two other historical cases regarding the effects on design collaboration due to the properties of design representations.

5.1 Flexible generic frameworks: some case histories

(Case 1) The funicular modelling revisited

In 1898, Eusebi Güell, a Catalonian textile industrialist, commissioned the design and construction of a new church to serve the industrial estate in Santa Coloma de Cervelló outside of Barcelona. The Colonia Güell Church project was then commissioned to the lead architect Antonio Gaudí. Working jointly with Gaudí on this project, there were the architect Jose Canaleta, working on the aspects of site planning and structural form, the structural engineer Eduardo Goetz, working on the area of building structural design and analysis, and the sculptor Juan Bertran, working on the aspect of architectural ornamentation (Martinell 1979, 335). A distinctive way of modelling the design of the church was developed to address the various aspects of the project.

- *The funicular modelling space.* At the inception of the project, an upside-down funicular model was constructed by the design group in a workshed (Figure 5.1). According to George R. Collins and Juan Bassegoda Nonell, this large three-dimensional funicular model was constructed with the following kinds of model-making objects and devices (Collins and Nonell 1983):
  - *Cords* hung in loops corresponding upside down to the placement and shapes of the piers and arches of the building's vault;
  - Pieces of irregularly shaped *Boards* attached onto the structural beams of the workshop representing contour lines and terraces of the building site;
  - *Weights* made of pellets contained in small sacks, corresponding to the distribution of loads of building components measured in the scale of 1/10,000, distorting the cords' catenary curves into funicular polygons when attached to the hung cords;

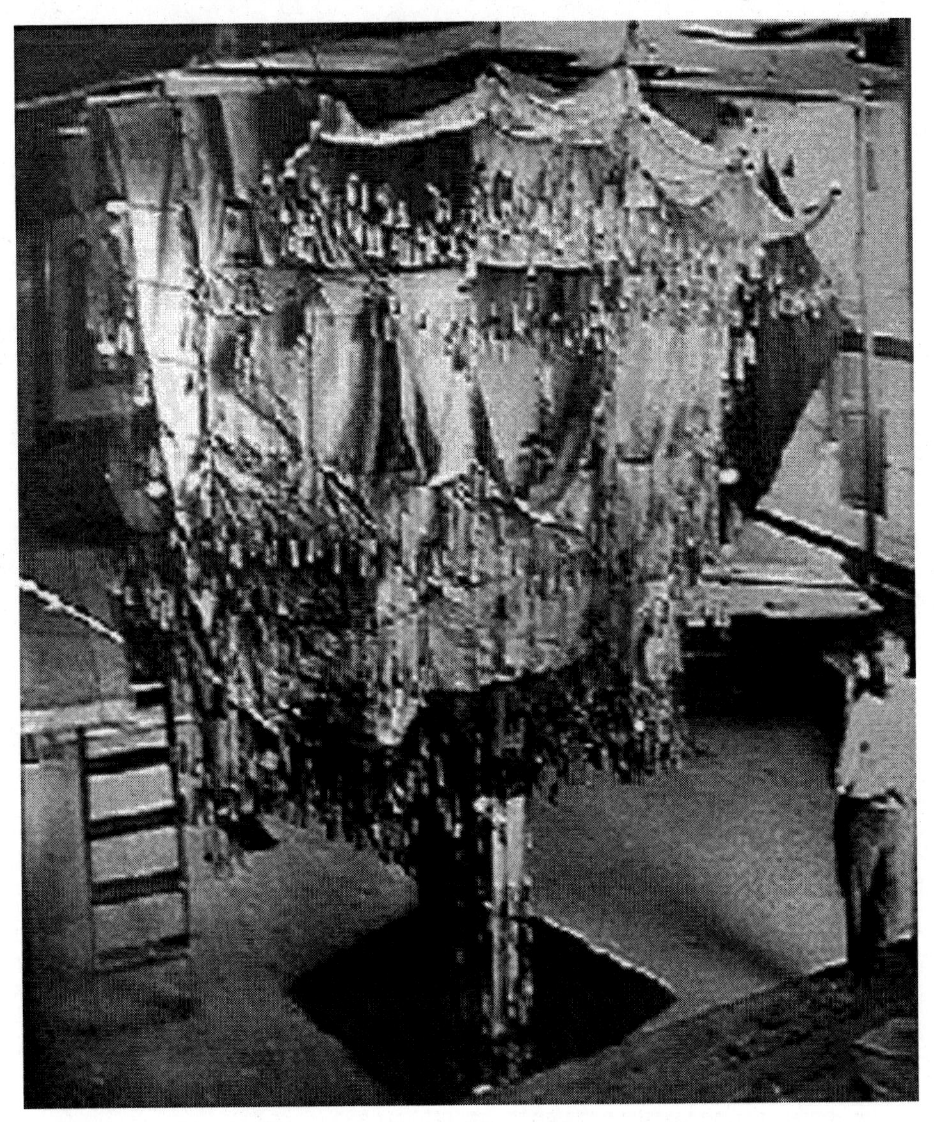

Figure 5.1 The three-dimensional funicular model constructed for the Colonia Güell Church project as it hung in the workshop. (Photo taken by the author from one of the historical pictures exhibited in the Gallery of Colonia Güell Church)

- *Fabric* draped onto the web of funicular polygon shaped cords, representing the volumetric effects of the building's exterior;
- A set of binding devices made of *Jointers*, *Hooks*, and *Clippers*, which do not represent any particular architectural elements of the building but allow designers to connect different kinds of model-making objects and manipulating parts of the funicular model. The jointers were used for attaching weights to cords; hooks for connecting the ends of cords to particular locations on the boards; clippers for clipping cords together at various heights (i.e., bifurcation).

- *Distributed domain design spaces.* The funicular model was shared and manipulated by all members of the group for performing various design tasks. Apart from the funicular model, there were other domain-oriented design approaches undertaken by different individuals according their design expertise. As documented in Collins and Nonell's account of the project history, the graphical evidence suggests the following activities:
  - The structural (civil) engineer's structural calculations: the distribution of loads in space and the thrusts of force lines were calculated by the engineer in a two-dimensional vector space; to him, the funicular model was seen as a three-dimensional illustration of planar and sectional *graphic static calculations* (Figure 5.2).
  - The architects' sketching of the exterior and interior spaces: photographs of the exterior and interior of the funicular model were taken and turned right-side up by the architects as the underlay information for modelling the locations, proportions, and shapes of opening (the fenestration of the building) (Figure 5.3 and Figure 5.4).
  - The sculptor's sketching out the ornamentation: the sculptor was concerned with the design of sculptural objects as the ornaments for the building's exterior and interior; like the architects, he was able to take photographs of the funicular models for his own design purposes and try out overlay sketching (Figure 5.5).
  - By juxtaposing the various sets of pictorial evidence, Figure 5.6 shows an overall view of the design collaboration involved in the Güell Church design: first, there is a common workspace used to construct and change the funicular structure; second, there is number of separate workspaces created and used by different participants for domain-specific design developments; and thirdly, the distributed modelling spaces are related to the funicular modelling space in one way or another.

- *Group interaction in and out of the funicular modelling space.* Given the above illustrations, I can make several points about what makes the funicular modelling space a shared workspace for the design team, and how the shared model serves as a medium of interaction among the participants:
  - First of all, the funicular modelling space was continuously developed and used by the design team for supporting long term participation; it was reported that the

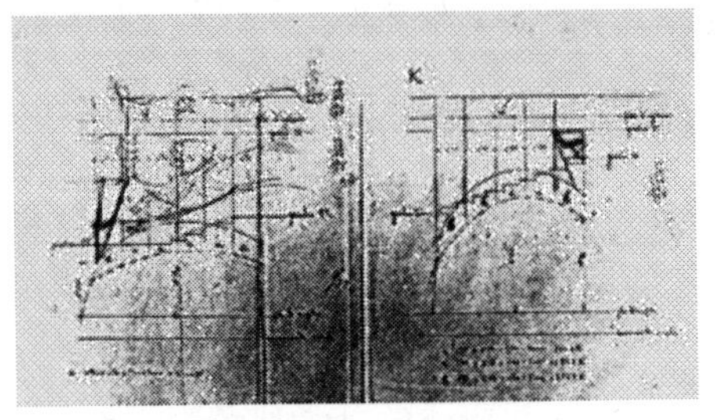

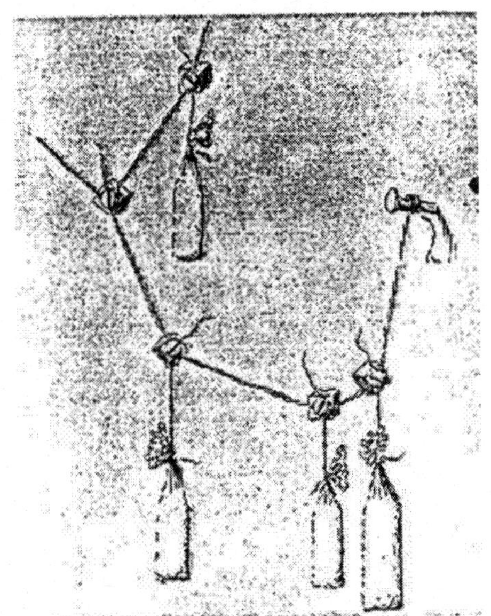

Figure 5.2 (top left and right) The structural engineer's approach to working out detailed structural calculations in relation to the funicular structure.

Figure 5.3 (bottom left and right) The architect's way of working out the exterior of the church in relation to the funicular structure.

Figure 5.4 *The architect's way of drawing out the interior of the church in relation to the funicular structure.*

Figure 5.5 *The sculptor's way of sketching of the interior ornamentation scheme in relation to the funicular structure.*

Figure 5.6 *An overview shows a number of distributed workspaces that participated in the Güell Church design project: (a) the funicular model constructed in the common workshop, (b) an inverted photograph taken inside the funicular structure on which the sculptor's ornamentation design was based, (c) the church's exterior design sketched out by the architects on top of inverted photographs taken outside the model, and (d) force lines constructed by the civil engineer on a projected elevation for structural calculations.*

participants collaborated on delicate exploratory work lasting over ten years[34] (Martinell 1979).

- Though all participants shared the same construction of a structural framework, the shared funicular modelling space allowed them to manipulate parts of the skeleton for reasons other than strictly structural. For instance, for the purpose of site planning, cords can be shifted to different hooks or by moving the hooks around the board; for modifying fenestration design, cords can be bifurcated at various heights by sliding the clippers along the force lines; for changing structural form, loads can be redistributed in space by controlling the number of pellets in sacks or by displacing the sacks' jointers to different positions on cords.
- For any state of the model, the participants could have individual interpretations and derive design information from, perhaps, different measurements; and the information derived further served as the basis for the designers to elaborate domain designs distributed among separate work settings.
- The fact that the earth's gravity was one of the (direct) forces in shaping the funicular model can explain how the group interaction could be coordinated by the shared modelling space. Through the action of Earth's gravitational force, the construction and manipulation of the model always conform to the mechanics of *funicular structure* (Schodek 1980). Therefore, a modelling action taken by a designer, for whatever reason, can trigger other designers' interpretations and evoke actions in response to the changing state of the funicular model.

The design and construction of Colonia Güell Church has often been seen as one of the most outstanding achievements in the works of Antonio Gaudí, though only the crypt part of the church was built as it stands today (Figure 5.7). The rich unique architectural expressions and the structural ingenuity of the building still captures enthusiasms among contemporary Gaudí scholars and design practitioners alike. Not to contend with the significance of Gaudí's leading the project development, I suggest that the case history can also be seen from the viewpoint of team working, which reveals a case of creative collaborative modelling of an innovative building design. What we can see is that revolving round the funicular model, the architects, structural engineers, and the sculptor had worked collaboratively on the aspects of architectural form, structural analysis, and interior ornamentation respectively for over a decade. The funicular structure is not a final direct representation of the architecture to be built; instead it is a kind of *intermediate* construct onto which different participants can carry out various design tasks. In many ways, parts of the structure can be substantiated with domain-specific design expressions as played out by different individuals of the design team.

We should also pay attention to the fact that parts of the funicular structure can be manipulated for various reasons other than solely a structural one; this was enabled by the various kinds of *connectors* built into the funicular structure. Members of the design team used these connectors not only to hook bits and pieces together but also to establish spatial relations among various parts of the funicular model. And the form of the funicular structure was constantly subject to participants' manipulation under the influence of the earth's gravitational force; any changes of the funicular form may mean

***Figure 5.7** The Colonia Güell Church as it stands today at Santa Coloma de Cervelló outside of Barcelona (photographed by the author).*

something as seen by some of the participants who interpreted the consequences of the change according to the domains they have been working in.

(Case 2) The Taxis Schemas

In the above I consider that the funicular model in the Colonia Güell Church design is an example of flexible generic structure that can function in a collaborative design setting. However, to enquire if the funicular structure is the only instance of flexible generic framework ever occurred in the history of architectural design, I have looked for other case histories. In the following, the second example of flexible generic frameworks is drawn from the Greek taxis schema and its descendants. The Greek taxis schema (or, the family of the taxis schemata) is basically a framework of "subdivision" in an architectural composition. Alexander Tzonis and Liane Lefaivre once characterised the taxis schema as follows (Tzonis and Lefaivre 1987, 9):

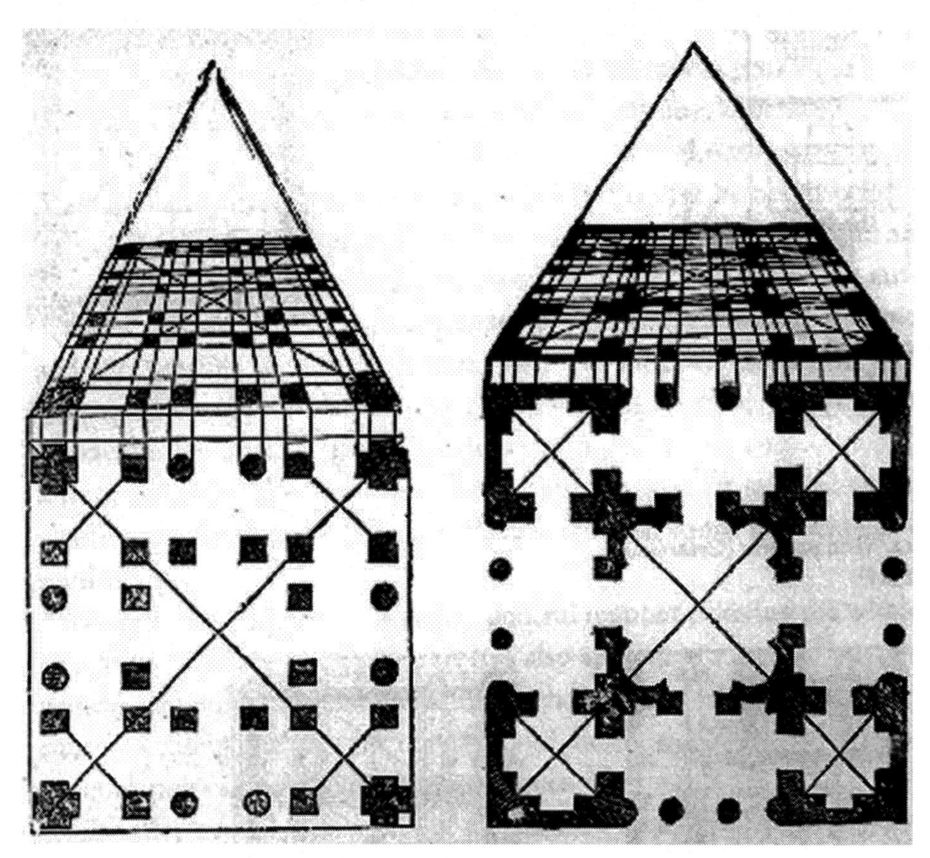

Figure 5.8 *An example of square taxis for a church design, attributed to Serlio, 1691.*

> Taxis divides a building into parts and fits into the resulting partitions the architectural elements, producing a coherent work. In other words, taxis constrains the placing of the architectural elements that populate a building by establishing successions of logically organized divisions of space.

Figure 5.8 shows an example of the "grid," or, more specifically, "square" schema. The square framework is presented in an overall, general manner encompassing, initially, the whole area of the building. With its vertical and horizontal lines set up, it can be applied in a more specific way in controlling the position of the element of wall, defining the space for the nave and the aisle. An even more elaborate architectural plan can be produced on the basis of the grid pattern.

Another taxis pattern is the *polar* schema. According to Tzonis and Lefaivre, a polar

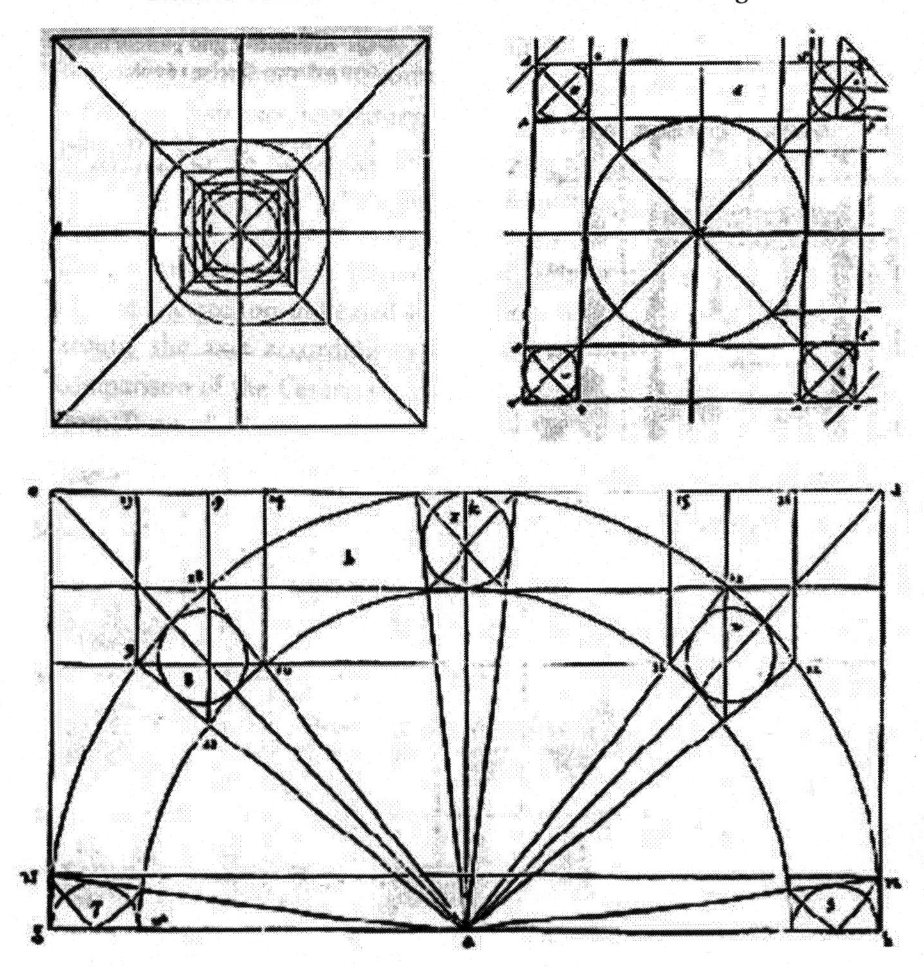

Figure 5.9 Examples of polar taxis schemata which partition building plans by means of contour and axis, attributed to Cousin, 1560.

taxis is a pattern where "One set of dividing lines forms concentric circles, while the other radiates from the common center of these circles." (Tzonis and Lefaivre 1987, 25). Figure 5.9 shows examples of polar taxis the spatial subdivision is running in a circular manner.

The Greek taxis frameworks have been continuously used in the classical architectural composition, and its influence on later work can be seen at least as late as Durand and Guadet (see Figures 5.10 and 5.11). As Tzonis and Lefaivre commented (Tzonis and Lefaivre 1987, 28):

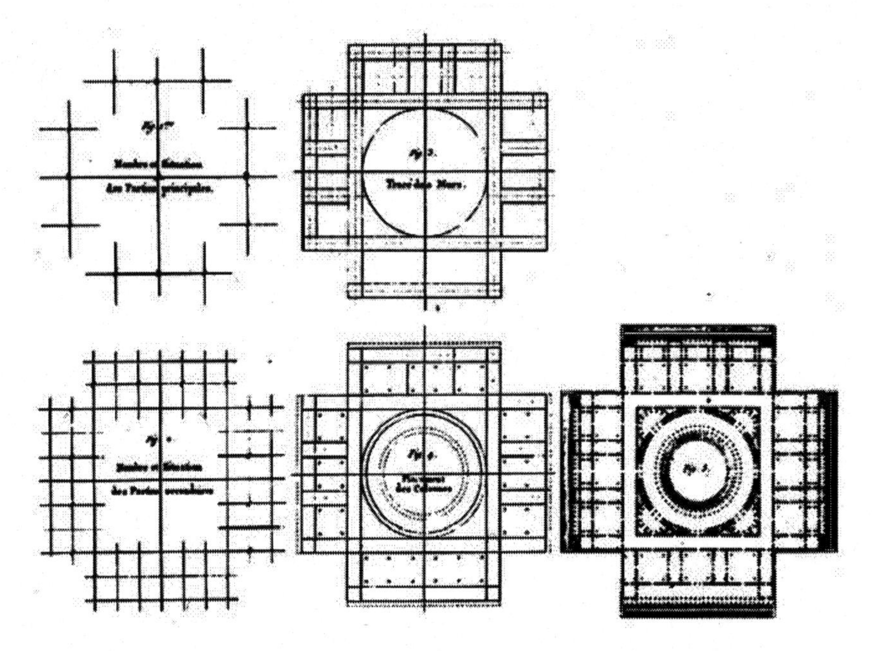

Figure 5.10 *Examples of taxis schemata developed in the nineteenth century showing plans with subdivisions of plans embedded in them, from Durand's Précis 1802-1805.*

> Toward the nineteenth century another way of specifying the divisions of a work became dominant: specification by an axis rather than by an outline. It is presupposed here that the architectural members of the section indicated by the axis are laid out–'balances'–around the axis according to bilateral symmetry.

What we see in the above examples of the taxis schemata is a family of frameworks or formal patterns that are subject to various systems of spatial or topological ordering and logic. Clearly, this is a different category of form-shaping force from that of the earth's gravitational force. Also, those lines of subdivision do not correspond directly to the buildings as designed but are operated as some controlling devices to locate and, thus, to interrelate clusters of building components. Although I did not obtain evidence that the taxis schemas and its descendants have been used in collaborative design processes, I may at least say that manipulation of those lines or grid patterns does not require particular domain-specific knowledge or skills. And it is not unreasonable to speculate that a spatial schema similar to the taxis ones can be operated by a design team among multiple layers of design spaces that correspond to various participating design domains.

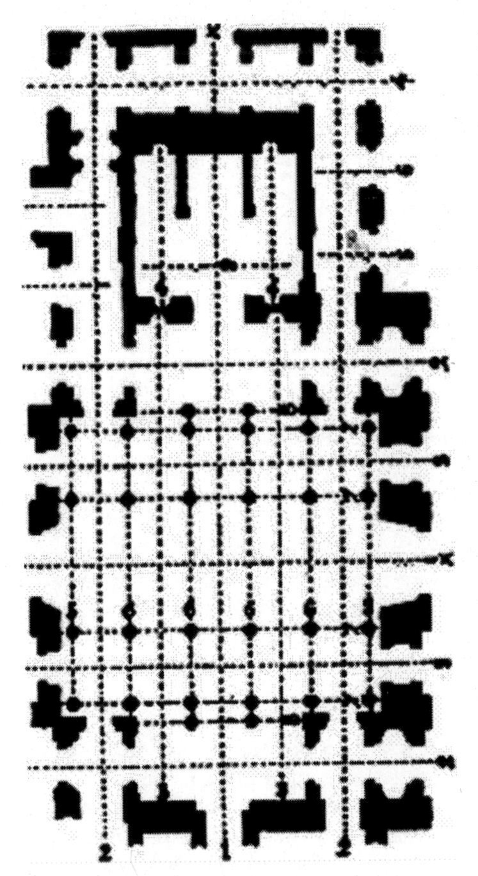

Figure 5.11 A grid pattern notation system showing the application of multiple taxis formulas on the same object by laying one over the other, from Guadet 1901-1904.

More importantly, based on the examples of the funicular modelling and the taxis schemas, we can conclude that there are at least two categories of *form-giving* forces operable in the field of building design: one can be characterised as *physical* (e.g., gravity, acoustics, light etc.), the other *intentional* (e.g., some humanly devised spatial/shape grammars or schemata). As shown, both categories can act as systems of formal constraints in the construction of structural or organisational frameworks.

5.2 From wholes to parts: A top-down scenario

I proposed a bottom-up scenario of collaborative design on the basis of the case studies presented in the preceding chapters. Similarly, drawing on the case studies of the funicular modelling and the taxis schemas seen above, I shall put forward another scenario of design collaboration. Mainly because of the uses of the funicular structures and the taxis schemas as common design development frameworks from the outset, a *top-down* scenario of collaborative design can be described concisely as follows:

The top-down scenario. At the inception of a design project, members of a design team work jointly in constructing a *group modelling space* for the modelling of a single structural or spatial framework as a *common generic structure*. There can be *forces* introduced in the modelling space with which specific forms or shapes of the structure can be evolved. When a participant applies projective (derivative) devices onto a state of a common generic structure, *derivative structures* can be produced and imported to his or her *individual modelling space* that is set up and used by the individual for domain design purpose.

By taking derivative structures as *design referents*, group members carry out separate strands of *domain design developments*; they may *substantiate* the imported derivative structures from parts of the generic structure with domain-specific constructs into more specific design expressions in their individual modelling spaces. In the course of elaborating domain design developments, some participants may be motivated, by seeing the design results in their working domains, to change parts of the derivative structures in use; to put the intended changes into effect, the individuals manipulate and modify the (corresponding) parts of the common generic structure. The changes thus proposed by one individual can subsequently cause further changes to be made in the derivative structures used by other participants.

The above scenario description is intended as an abstraction of the similarities of the case histories. I consider it possible to give a systematic account of the top-down scenario of collaborative design based on the properties observed in flexible generic structures. First, these generic frameworks are design representations capable of facilitating multiple design perspectives in terms of model construction, manipulation, and substantiation. Secondly, these structural objects (two or three-dimensional) themselves can be expressions of formal relations that govern how parts are interrelated with one another with reference to the structural wholes. In contrast with the bottom-up approach, here we have *emerging parts* instead of an emerging whole. The designing of parts in the top-down approach is distributed among members' individual workspaces in accordance with a representation of architectural whole (i.e., a common flexible generic structure) constructed in a group workspace.

In the top-down scenario, the judging of how well a project design as a whole is progressing is based on how parts are being developed at each of the design domains. Members of the design team may actually see their present domain design works as the *consequences* of the current state of a common generic framework. A generic structure can be amended by team members who feel it is inadequate or unsatisfactory, hence being *multi-viewpoint*. However, any changes made by one participant onto a generic framework may give rise to problems for other participants who share the framework in developing his or her domain designs. This is due to the actions of a form-giving force which constantly shapes the state of a structure until a *balanced* or *equilibrium* state is reached. Therefore, we can say that the sharing of a common generic structure gives rise to group interaction in collaborative design. But what is involved for members of a design team to share flexible generic structures? Let us consider the following more specific questions:

1 What properties can we attribute to the representational elements or devices that enable group members to construct, manipulate and change a common generic structure from multiple perspectives?
2 What are the conditions for a generic structure constructed in a group workspace to be *shareable* among the participating designers? What are the constraints on the *shareability* of a generic framework?
3 How is the development of common generic structures participated by group members interrelated with the developments of domain-oriented designs as separate individual design responsibilities?
4 Given the parallel engagement in group and in individual modelling activities, what do members of a design team have to know and perform in order to initiate and sustain group communications for continuous design development?

To answer the questions raised above thoroughly, I shall adopt again the method of situation-theoretical analysis in the next section.

5.3 Top-down flow of information in collaborative design

The situation-theoretical analysis of the top-down scenario consists of the following aspects: an action-space matrix, an enumeration of the situation types, and then a schematic view of the flow of information. Like before, the aim is to elicit the constraints on the flow of information among the situation types identified in the top-down scenario. Following this, the core issues and requirements for developing computer support are discussed.

Modelling spaces and modelling acts

Regarding the types of modelling spaces, the top-down pattern is the same as the bottom-up one. But the entities that constitute the modelling spaces in the top-down case contain some differences:

Group Modelling Space (GMS). The term *group modelling space* refers to a modelling space created and utilised jointly by designers participating in a design project. In the top-down scenario, one of the key functions of a GMS is its use by all participants in modelling some kind of flexible generic structures (see more details below), and it can be defined in terms of the following components:

- *Model constructs*—A collection of elementary objects that participants introduce to be used in the construction of a flexible generic structure whose state is constantly subject to the form-giving forces deployed.
- *Form-giving forces*—Fields of physical or other kinds of forces or constraints that participants deploy to shape or deform parts of a generic structure. Given a force field introduced in a group modelling space, the state of flexible generic structure is of a common concern to members of a design team; a state of *equilibrium* or *satisfaction* as manifested in a configuration of instances of model constructs in reaction to the forces or constraints applied.
- *Manipulative operations*—Operations that enable participants to displace, transpose, or aggregate instances of various types of model constructs such that flexible generic structures can be created and changed.
- *Derivative operations*—Operations that allow participants to perform certain spatial actions, such as sectioning, projecting, tracing, truncating and mapping, so that *derivative structures* can be generated.

Individual Modelling Spaces (IMS). Individual modelling spaces are created and used by individuals for the development of domain design works. In general, to carry out domain design tasks, members of a design team may set up their IMSs in accordance with *individual object world*—a designer's design world, consisting of (domain-specific) notations and tools for coding, visualising, manipulating, and evaluating design expressions. The same set of modelling acts seen in the bottom-up scenario can be applied here, i.e., *representation*, *generation*, *interpretation*, and *modification*. Since there are

Modelling Acts / Modelling Spaces	*Abstraction*	*Generation*	*Interpretation*	*Modeification*
IMSs	***Individual Object World (IOW)***	***Domain Design Expressions (DDE)***	***Local Design Assessment (LDA)***	***Changes in DDE or in IOW***
GMS	***Shared Construction Set (SCS)***	***Common Generic Structures (CGS)***	***Derivative Sturctures (DS)***	***Changes in CGS or in SCS***

***Table 5.1** A space-act matrix specifying eight design states in the top-down scenario of collaborative design.*

no fundamental differences in the acts of design modelling, the previous descriptions of these acts presented in Chapter 3 can also be considered adequate in the top-down scenario.

A Space-Act Matrix. Given the modelling spaces and modelling acts outlined above, a space-act matrix can now be constructed (see Table 5.1). Like before, a space-act matrix is constructed to specify the range of design states (i.e., the information carriers in the situation-theoretical terms), which, in turn, gives rise to a listing of the situation types in the top-down scenario, serving as the major headings for a situation-theoretical exposition.

The situation types in collaborative design involving flexible generic structures
Recall that a situation type is an abstract description of some natural activities that allow for a common "type abstraction" as stipulated in the theory of situations. Information can flow from one situation to another under some systematic relations (constraints) that exist between the situation types to which the two situations belong. I suggested in Chapter 3 that a situation type in collaborative design could be defined in terms of generating a design state by the performing of a modelling act in a modelling space. By matching triples that consist of various modelling acts, spaces and design states, a set of situation types in collaborative design can be enumerated. As a result of following the situation-theoretical way of working, we now arrive at eight situation types, which characterise some if not all aspects of the top-down approach to collaborative design.

1 [*representation*, GMS, **Shared Construction Set**]—The type of situations in which members of a design team perform *design representations* in a group modelling space and results in a *shared construction set*.

An instance of the situation type can be seen in the funicular modelling case study where the basic model constructs (e.g., board, cord, weights, etc.) for building funicular structures are introduced in the workshop. The range of model constructs is the participants' abstractions of the design elements or the key factors concerning the building project. I shall term the set of model constructs introduced in a shared workspace for modelling generic structures (see below) as *shared construction set*. Clearly, a shared construction set is an information carrier at a very low level that contains information regarding where the design modelling starts. For the present situation type, we can even extend the scope of abstraction to include the introduction of form-giving forces. In most cases, designers may be not necessarily involved in representing a form-giving force to their common interest in a group modelling space. As in the funicular modelling case, designers simply make conscious uses of the gravitational force that is readily available in the workspace. However, if we consider design modelling to take place in a digital world, the representation of a system of form-giving forces can be of no small matter.

2 [*representation*, IMSs, **Individual Object World**]—The type of situations in which an individual sets up a scheme of design representation in his or her own modelling space and results in an individual object world.

Similarly to the bottom-up pattern, the term "individual object world" is used to denote this type of design state. This is the type of situation where elements of design representation are introduced by a designer working in a particular domain of a design project. A designer's abstraction of an individual object world can be seen as the point where an individual's design expertise starts to matter. A designer's object world is the basis for his or her substantiating parts of common generic structures with domain-specific design expressions (see below). Like the state of a shared construction set, an individual object world is also a low-level information carrier subject to changes whenever intended by its designer/author.

3 [*generation*, GMS, **Common Generic Structures**]—The type of situations whereby one or more designers perform design generation in a group modelling space resulting in a state of common generic structure.

Given a state of shared construction set is made available in a group modelling space, this is the situation in which group members work concurrently in generating a configuration of generic flexible structure, which contains 2-D or 3-D generic objects, representing, mainly, a kind of spatial framework or skeleton. As one of the salient features of the top-down approach, the framework is constructed and can be subsequently used by all participants working in different domains of a design project. Drawing on the instances from the case histories, the following general properties can be said about a flexible generic structure:

- *Deformable*. A common generic structure is made of instances of model constructs that are configured in a field of physical forces or formal constraints. Being constructed and shared by all participants in a group modelling space, a structure of this nature is meaningful and useful if it reveals some resultant spatial forms or geometrical shapes. Changes in forces or constraints applied or in the attributes of deployed model constructs may *deform* a structure into different states. The

deformability entails that the construction of the generic structure is subject to the behaviours of certain systems of constraint satisfaction or equilibrium of forces.
- *Multiple-viewpoint*. Parts of a generic structure can be manipulated by participants from multiple points of view for different reasons. The multiplicity is firstly achieved by participants introducing types of model constructs that correspond to various *perspectives* concerning a design modelling project (e.g. site, structure, enclosure, opening, etc., in building design). Secondly, there are multiple ways allowed to assemble or detach model constructs while modifying parts of the generic structure. This multiplicity lies in a range of *connecting devices* that can be used to bind instances of various types of model constructs in the course of modelling.
- *Derivable*. A state of generic structure can be applied with *derivative devices* as intended by any individual designers. Applied derivations can generate instances of derivative structures (see below) that can be further transferred to individual workspaces for domain uses. The derivability allows participants to establish *referencing* relations between individual design developments and the evolution in their sharing a generic structure. Derivative operations may be developed as a part of a shared construction set.

4 [*generation*, IMSs, **Domain Design Expressions**]—The type of situations where an individual performs design generation in his or her modelling space resulting in domain design expressions.

Given an individual object world made available in an individual modelling space, this is the situation where a designer generates design expressions specific to his or her domain design tasks at hand. The sharing of a common generic structure in a way is only half of the top-down approach. As a project develops, final design products often go well beyond the construction of generic skeletons. An equally important development is concerned with how the generic structures can be substantiated with more specific design contents into domain design expressions. In the top-down scenario, domain design expressions are the outcomes from participants specialising parts of common generic structures based on their domain design expertise.

Being different from the bottom-up, the generation of domain design solutions here is closely associated with another type of design state–derivative structures (see below). The difference lies in that all domain design expressions are generated on the basis of the information derived from a state of common generic structure. That is to say, designers generate design expressions by taking derived structures as underlying design references.

5 [*interpretation*, GMS, **Derivative Structures**]—The type of situations where a designer's interpretation of a state of common generic structure in a group modelling space leads to *derivative structures* for domain design purposes.

Given derivative operations defined in a group modelling space, this is the situation where a designer produces information of a secondary order in contrast with a state of common generic structure as the first order (for instance, the inverted photographs taken from the exterior or interior of the funicular model). As we may observe from the funicular modelling case, *derivative structures* are basically 2D or 3D pictorial objects representing some static spatial frames or skeletons. Images contained in a derivative structure, once imported into individual workspaces, can serve the individuals as design referents in generating domain design expressions. I consider this activity as an interpretation because how and where to apply derivative operations involves a designer's opinions or instructions of what and how to carry domain design tasks in relation to the state of a shared generic structure. Therefore, this is similar to say that to designers with different design interests and responsibilities, a common generic structure might have different meanings to them as to how it might be used for the purpose of developing domain designs.

To look at derivative structures further, in the course of exploring possibilities of domain designs, a designer may feel the need to manipulate parts of the underlying design referents. However, the manipulation has to be largely indirect, because instances of derivative structures can be basically *frozen* images on the whole, and they can only be manipulated not in parts but as a whole. To effect intended changes in the referents, changes have to be made in the common generic structure first, so that a new state of derivative structures containing the intended changes can be reinterpreted. This is like taking a new photo of an altered state of the funicular model to be used in trying out new schemes of the church's exterior or interior design.

6 [*interpretation*, IMSs, **Local Design Assessment**]—The type of situations where a designer performing design interpretation in his or her own modelling space results in local design assessment.

When domain design expressions are made in individual modelling spaces, this is the situation where the designers give interpretations of the expressions produced and arrive at some assessments from domain-specific viewpoints. Since each piece of domain design expression is generated with reference to a state of derivative structure, the resultant expression, as viewed and assessed by its author from a particular domain perspective, can be interpreted as a *design consequence* of common generic structure. A participant may be motivated by his or her state of local design assessment to carry out design modification in an individual or group modelling space.

7 [*modification*, GMS, **Changes in CGS and/or SCS**]—The type of situations where one or more designers performing design modification in a group modelling space results in changes to be made in common generic structures and/or in the shared construction set.

When a local design assessment arises in the mind of a participant, he or she will get access to the common generic structure and change the attributes of model constructs (for instance, moving a cord from one place to another, or distributing weights in a different manner as seen in the funicular modelling case), so that undesirable design consequences as manifested in her local design expressions can be removed. Or, at a more fundamental level, some members of the design team may introduce new model constructs to the existing model construction set, so that a new aspect of modelling common generic structures can be opened up. In either case, the type of situations will give rise to communication and coordination among group members. Details of the group communication involved will follow in the next chapter on the constraint analysis of information flow.

8 [*modification*, IMSs, **Changes in DDE and/or IOW**]—The type of situations in which a designer performing design modification in his or her own modelling space results in changes in domain design expressions and/or in the individual object world.

Given that a new state of derivative structure is imported in a local modelling space, this is the situation where the designer executes intended changes in parts of a design expression in question. As newer design expressions are produced with reference to the newly underlay derivative structure, the designer may go through further interpretations and produce more local design assessments. So the top-down approach to group design goes on until, perhaps, no participants in the group express further needs to change any parts of either the common generic structure or domain design expressions.

We may now acquire a chart of information flow by putting the above situation-theoretical types into an overall picture that reflects the features described in the scenario. To better illustrate the information flow, a diagram showing a scheme of the information flow is presented in Figure 5.12. Assuming a design team consisting of three designers working in various aspects of a building project, the chart starts with the formation of a shared construction set upon which the generation of common generic structures is based. Derivative structures are derived from the generic structure and imported to a number of distributed individual modelling spaces. With reference to the underlying derivative structures imported, domain design expressions are made in individual object worlds. Domain-oriented interpretations of the expressions result in local design judgements which, in turn, may motivate repeated replacement of the underlying derivative structures. The design changes thus initiated in each domain may result in more changes to be made in the common generic structure. The diagram provides a basis for a systematic survey of the constraints on the information flow among the situation types in the top-down scenario of collaborative design, which will be explained further in the next chapter.

5.4 Regarding the Dutch architectural Structuralism

The pattern of team working as studied in this chapter is an approach to collaborative building design that demonstrates some features of group dynamism that is distinctive from the bottom-up approach described in Chapters 3 and 4. The term "flexible generic structures" has been used extensively in this chapter to describe a different type of collaborative design modelling process. I suppose that my use of the term comes firstly from seeing the funicular model constructed by the Colonia Güell Church project. This particular case shows a striking array of design representations and expressions that

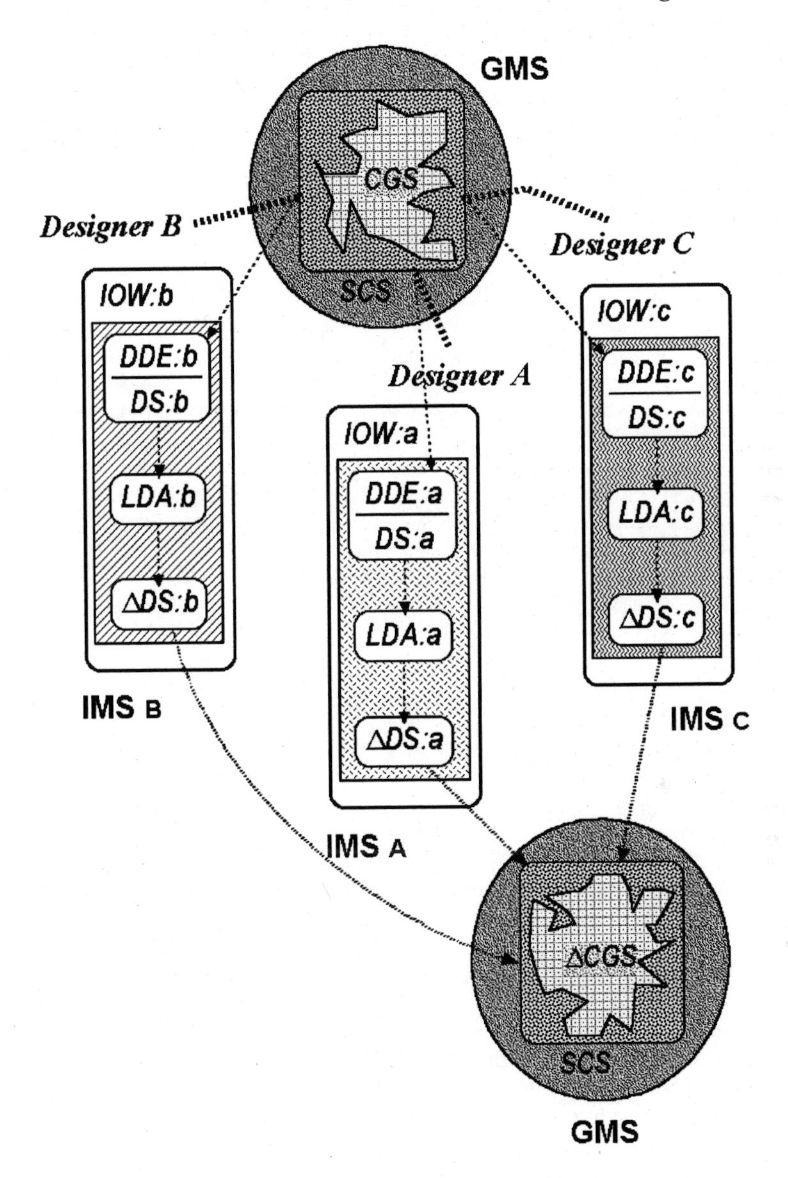

Figure 5.12 A scheme of the flow of information among the situation types classified in the top-down approach. (Note that the number of designers shown on this scheme is only an assumption; in theory, the number can be scaled up to any group size.)

correspond to the collaborative building design processes. In relation to team working, this particular kind of representational object is created collectively from the outset but can be viewed and substantiated by members of a design team from different design perspectives.

As a matter of drawing out more related literature, the "structures" used in describing the top-down scenario of design collaboration bears some connection with the structures that are of interests to a group of Dutch architects and theoreticians, including Aldo van Eyck, Jaap Bakema and Herman Hertzberger among others, who published writings and projects about *architectural Structuralism* in the late 1950s in the Dutch journal *Forum*. In his more recent discourse on architectural Structuralism, Herman Hertzberger, an influential contemporary Dutch architect and theorist, has the following statements regarding "structures" (Hertzberger 1991, 94 and 103):

> Broadly speaking, 'structure' stands for the collective, general, (more) objective, and permits interpretation in terms of what is expected and demanded of it in a specific situation. One could also speak of structure in connection with a building or an urban plan: a large form which, changing little or not at all, is suitable and adequate for accommodating different situations because it offers fresh opportunities time and again for new uses.

> The main form which we called structure is collective by nature, it is usually controlled by a governing body, and is essentially public. Control over the uses to which it is put ranges from more public to more private, depending on the commercial interests involved.

In Hertzberger's view, for instance, the canals in Amsterdam, the viaduct on the Rue Rambouillet in Paris are good examples of structures that are "always ready to accommodate new purposes which in their turns add new meanings to the surroundings." In architectural Structuralism, structures are some man-made environmental artefacts that are general kind of constructions and belong to public domains. These structures as we see them today may have been brought into existence by some communities many generations before, yet they have been allowing interpretations of how the structures may be adapted or reused by later generations. In comparison, the structures in the top-down approach to collaborative design as shown in this chapter bear a similar connotation but in a much restricted world of applications. Generic structures are representations in architectural modelling, which are "public" to the members of a design team (or, public to the domains of a building design project) and any parts of a structure can be interpreted differently by designers working in different domains. The property of a common structure that allows for multiple individual interpretations, rather than the other way around, has been put forth quite strongly by Hertzberger as follows (Hertzberger 1991, 147):

> Collective interpretations of individual living patterns must be abandoned. What we need is a diversity of space in which the different functions can be sublimated to become archetypal forms, which make individual interpretation of the communal living-pattern

> possible by virtue of their ability to accommodate and absorb, and indeed to induce every desired function and alteration thereof.

Regarding changes and evolutions of structures in architectural Structuralism, Hertzberger remarks on the idea of "polyvalent form" in the following (Hertzberger 1991, 147):

> The only constructive approach to a situation that is subject to change is a form that starts out from this changefulness as a permanent—that is, essentially a static—given factor: a form which is polyvalent. In other words, a form that can be put to different uses without having to undergo changes itself, so that a minimal flexibility can still produce an optimal solution.

From a design point of view, the funicular structure in the Colonia Güell Church project and the examples of the taxis schemas are polyvalent representations upon which multiple layers of constructions in different domains of design can be attached. And I think that "a form can accommodate different uses without undergoing changes itself" can well be just a shorthand for perhaps in actuality a complex and dynamic process in which a form can always reach a state of *equilibrium* under the influences of a system of form-shaping forces in action. The fact that the funicular model has been "tinkered" by its designers for more than ten years in the design of the church without undergoing fundamental changes itself may be seen as essentially static.

Summary

In the search for evidence of collaborative design we come across a set of case histories that present different features of teamwork in architectural design from the ones observed in the bottom-up pattern. In this chapter, the case study on the funicular modelling in the Colonia Güell Church project shows that some kind of flexible generic framework can be employed by a design team to sustain group dynamics through a project's lifetime. The Taxis Schemas as a spatial conceptual device for subdividing architectural spaces provide further examples of flexible generic structure. There are important properties of the shared structures that enable group interaction throughout the design development process: genericity, flexibility, and multiple-viewpoint. Direct or indirect communication among team members can arise from different uses of the common structures. Based on the case histories, a stop-down scenario of collaborative design is presented. Applying again the situation-theoretical framework introduced in Chapter 3, an exposition of the top-down pattern of collaborative design is given by listing and explaining the eight situation types derived from the space-act matrix and the design states. The top-down pattern of collaborative design bears no direct relations with the Structuralism in linguistics, anthropology, and psychoanalysis. However, the structures as meant by the Dutch architectural Structuralism do have some correspondences to the representational properties of flexible generic structures seen in collaborative modelling of building design. The Dutch architect Herman Hertzberger spoke of structure in a building or an urban plan as a large collective, general and objective form which is polyvalent and is capable of accommodating new uses (interpretations) as situations arise.

6 Joint Substantiation of Common Generic Structures

The construction and sharing of flexible generic structures is one of the most salient features observed in an approach to team working in design which I have described as the top-down scenario of collaborative design. Working in this approach, members of a design team have two roles to play in parallel: one is to take part in the construction of a generic structure, and the other is to substantiate parts of the common structure with domain-specific design expressions. In the preceding chapter, I gave a situation-theoretical exposition regarding the situation types in the scenario. Based on the intermediate results, I shall continue to discuss the logic of the top-down pattern of team working by eliciting the constraints on the flow of information among the situation types listed and explained earlier.

The current set of constraints identified in turn suggests a number of issues to be considered if computer supports are to be developed in support of aspects of the top-down approach to collaborative design. Regarding the issue of computer-based representation and modelling of flexible generic structures, which is one of the most challenging areas in my view, a review of related research done in computer graphics and artificial intelligence is presented. In particular, the computational techniques of constraint-based object-oriented graphics are potentially applicable to the kind of graphic modelling environment required. I shall then propose a system framework for building a constraint-based object-oriented graphic modelling and mediating system in support of multidisciplinary collaborative synthesis of built form. Finally, to round off my current exposition of collaborative design, I shall discuss more about the contrasts between the bottom-up and top-down scenarios of team working to highlight some of the findings presented in the current design studies.

6.1 Constraints on sharing flexible generic frameworks

If we look again at the diagram of information flow in Figure 5.12, we can identify two main areas of flow in which group interaction is necessarily involved: from *shared construction set* to *common generic structure*, and from *changes in derivative structure* to *changes in common generic structure*. For each of the two areas a detailed description and explanation is provided in the following. There are another four possible strands of information flow as shown in the diagram, namely, from *common generic structure* to *derivative structure*, from derivative structure to domain design expression, from domain design expression to local design assessment, and from local design assessment to changes in derivative structure. These flows mainly take place in individual modelling spaces without involving communication among group members. The constraints on these types of flow are therefore mainly to do with requirements for supporting individual rather than collaborative design work.

1 [*representation*, GMS, **Shared Construction Set**] → [*generation*, GMS, **Common Generic Structures**]
Information flows from a situation type where a shared construction set is set up in a group modelling space to a situation type where a common generic structure is generated in a group modelling space.

Question: *Given a shared construction set available in a group modelling space, what might make it impossible that a configuration of a generic structure becomes "common" or "shareable" among all members of a design team?*

An important issue here is how to achieve and evolve a common framework of project work that can serve the participants to develop areas of domain-specific design substantiation. The status of being *shareable* of a common generic structure can be critical to the continuing of teamwork, since the structure is used by participants as underlying frameworks for domain design developments. The "shareability" of a generic structure therefore reflects the status of common understanding and judgement upheld by the team members. Under certain circumstances, however, a flexible generic structure constructed may cease to be shareable, hence, design collaboration cannot continue. Consider the following conditions:

- **Deformable.** A generic structure becomes not deformable in a group modelling space due to a state of *equilibrium* of the structure cannot be reached or sustained. For instance, a generic structure is torn apart or collapses under a state of form-shaping forces applied.
- **Multiple-viewpoint.** In the course of modelling because of the lack of certain types of model constructs or connectors, some members of the design team may not get access to and manipulate certain parts of the current state of the generic structure upon which their domain design developments depend.
- **Generic.** The state of a generic structure becomes unusable to some designers because its derivative structures are not generic of an appropriate level to serve the purposes of domain-oriented design substantiation.

For a generic structure to be shareable, team members must ensure that the requirements of maintaining the structure's states of being *deformable*, *generic*, and *multi-viewpoint* are satisfied. But how would a design team manage to do so? If we are to weigh the three factors above, the availability of proper *model connectors* and *model constructs* in relation to a system of form-giving forces is perhaps the most crucial one. Types of model constructs may be brought into project work by participants on a trial basis in the first instance. Naturally, if types of model constructs are introduced by different designers, to make use of instances of the constructs in building a flexible intermediate common framework, a way of connecting or binding various types of constructs must be found. This can be design experiment jointly carried out by the contributing designers. A collection of model connectors is a kind of connective device that can be used to join instances of different types of model constructs in relation to a field of form-giving forces in action. Here I am thinking of the kind of objects like *hooks*, *clippers*, *jointers*, etc., as seen in the funicular modelling case. Once a set of connective devices is made available in a group modelling space, the properties of being deformable, multi-viewpoint, and generic of a structure can then be built up and tested.

Deformability of a generic structure may be lost if an equilibrium state of the structure cannot be restored regardless of how form-shaping forces form—governing rules may be applied. If the state of a generic structure is no longer deformable, members of the design team are not in a position to manipulate the generic structure in any meaningful ways. An irretrievable loss of deformability can be caused by the failures of some instances of model constructs or connectors due to the generic structure's inability to withstand the applied forces or rules. We can easily imagine such possible failures in the modelling of the funicular structure. In the case of the taxis schemas, this is like that no further rules can applied to change the state of a spatial configuration. However, it should be noted that a breakdown of equilibrium does not necessarily mean a (disastrous) end of group design; in a way, the design team may thus realise the limit of their current construction with regard to the forces applied. To push the boundary further, the design team may devise new types of model constructs/connectors, or alternative ways of manipulating form-shaping forces in, perhaps, a more controlled or incremental manner. I shall use the following expression to denote our first constraint on information flow in the top-down pattern.

Constraint 1: (***SCS*** → ***CGS***) ⇒ Adequate set of model constructs and connectors, and deformability preserving in manipulating form-shaping forces or rules

2 [*modification*, IMSs, **Changes in Derivative Structures**] → [*modification*, GMS, **Changes in Common Generic Structures**]

Information flows from a situation where one or more participants put up proposals for design changes to be made in the derivative structures under their uses to a situation where design changes are to be made in parts of a generic structure.

Question: *What is involved for designers to make changes in their domain design expressions in relation to the state of a common generic structure?*

Designers working in domain-specific aspects of project work do not develop domain designs from scratch; they generate design parts on the basis of the individual derivative structures taken separately from a state of common generic structure shared by the design team. Consequently, making changes in domain design expressions has to take into account of the underlying derivative structures which in turn may have implications of amending the state of a common generic structure. Seen from a teamwork point of view, it is more appropriate to address changes in derivative structures rather than in domain design expressions. However, to better address the question raised above, let us look more closely at the *consistency* relations among the design states of common generic structure, derivative structure, and domain design expression.

(a) R_d[***CGS***, ***DS***]—Derivative structures are said to be *derived from* a state of a common

generic structure. Therefore, a state of CGS always stands in a relation, denoted as R_d, to a state of DS. In theory, the type of relation R_d can be characterised in terms of the derivative devices or methods employed and the spatio-temporal locations (relative to the state of a **CGS**) of executing the derivative operations.

(b) R_f[**DS, DDE**]—Domain design expressions are constructed by participants *with reference to* underlying derivative structures. Therefore, a state of **DS** always stands in a relation denoted as R_f, to a state of **DDE**. Examples of the type of referencing relation R_f are geometrical or spatial referencing and domain design contents substantiating.

(c) R_d[**CGS, DS**] • R_f[**DS, DDE**]—By compounding the above two relations, a more complex relation among states of common generic structure, derivative structure and domain design expression can be formulated. This compounded relation indicates what is involved in maintaining consistency of design information or data in two aspects:

CGS → **CGS'** (1.a): some participants making changes in a state of **CGS**;
DS → **DS'** (2.a): changes in a state of derivative structure are implied due to the R_d relation between **DS** and **CGS**;
DDE → **DDE'** (3.a): changes in a state of domain design expression are implied with reference to the R_f relation between **DS** and **DDE**.

Or the information flow can be the other way around as follows:

DDE → **DDE'** (1.b): some participants making changes in a state of his or her **DDE**;
DS → **DS'** (2.b): changes in a state of derivative structure are implied due to a R_f relation between **DS** and **DDE**;
CGS → **CGS'** (3.b): changes in a state of common generic structure are implied due to a R_d relation between **DS** and **CGS**.

Based on the consistency relations among different design states shown above, two other constraints on collaboration can be established as follows.

Suppose, at some design stage, a participant (say, Designer A) decides to make some changes in DS_A (to denote the set of derivative structures used by Designer A) to maintain (or validate) an intended domain design decision. Consequently, A's changing DS_A leads to a changing state of the common generic structure, which is shared by other participants (say, Designers B and C). Assuming that the deformability of the **CGS** is intact, the derivative structures, DS_B and DS_C as used by B and C respectively, may be changed in order to maintain the derivative relations introduced earlier. This gives rise to at least two circumstances calling for communication among Designers A, B, and C:

Constraint 2: $[(DS_A \rightarrow DS'_A), (DS_B \rightarrow DS'_B), (DS_C \rightarrow DS'_C)] \rightarrow [CGS \rightarrow CGS'] \Rightarrow$ **Coordination**

Coordination is involved if A's changing DS_A is seen and judged desirable by B and C as they inspect the consequent states of their own derivative structures. Under this circumstance, B and C need to coordinate A's design changes by making changes or developing new elements in their domain design works in respect of the changed DS_B and DS_C.

Constraint 3: $[(DS_A \rightarrow DS'_A), (DS_B \rightarrow DS'_B), (DS_C \rightarrow DS'_C)] \rightarrow [CGS \rightarrow CGS'] \Rightarrow$ **Negotiation**

Negotiation is involved if A's changing DS_A is judged undesirable or unacceptable by B and/or C as they inspect the differences occurred in the changing states of DS_B and/or DS_C. Under this circumstance, Designer A needs to negotiate with B and/or C by either dropping completely the intended changes in DS_A, where A faces with developing his or her domain designs in a different direction, or requesting B's and/or C's suggestions of the extent to which the changes in DS_A are acceptable. A recognising suggestions from B and/or C is the same as B and/or C recognising A's intention of changing DS_A as reflected in the state of common generic structure; the cognitive basis for A to do so is the deformability of CGS and the relations between CGS and DS.

According to the constraints elicited above, the logic of collaborative design in the top-down scenario can be described more holistically as follows: participants in a design project collaborate with one another by building a generic structure in a shared work space, which provides a common framework capable of linking domain design developments undertaken by participants in a distributed manner. A deformable and generic structure can be built and evolved from multiple viewpoints, if and only if an adequate set of *model constructs* and *connectors* is made available. *Coordination* is involved if one designer's intention to change parts of a generic structure is acceptable to other team members; otherwise *negotiation* among participants for making design changes is necessarily involved until the *shareability* of the generic structure is regained.

6.2 Issues in supporting joint design substantiation

Like the investigation of the bottom-up requirements in the previous chapters, a formal and complete requirement specification of a computational system aimed at the top-down approach to teamwork is not the goal here. Instead, we shall look at areas of developing computer support which have not yet been fully addressed by the earlier collaborative drawing support tools. It has to be said, the current study does not suggest that a direct imitation of the bottom-up or top-down approach may lead to a fruitful development of computer-supported collaborative design. Nor do I consider it particularly desirable in practice to give designers a computer-based environment that implements exactly what we see from the various case studies. Any attempts of doing so can end up with some prescriptive systems that may actually stifle creative and dynamic design collaboration. A more worthwhile direction is to develop general computational representation and communication mechanisms that can be tailored by design teams in response to the emergent group dynamics in design integration and distribution. Following the constraints on collaboration arrived at the above, I shall move onto discussing the issues of supporting the representation and communication requirements elicited in the top-down scenario.

Support for representation in group modelling space

Collaborative design begins with team members forming a shared approach to design modelling in a group workspace. A computer-based collaborative design environment can be devised to facilitate the group activity of joint abstraction by providing, first of all, abstract generic constructs with which design participants can build up their own modelling constructs pertinent to the project at hand. Without further research, however, it is not possible to suggest specifically what basic system constructs should be included in a system implementation. It may well be an open-ended process of development. Based on the current analysis of the top-down approach, the provision of system support for building up shared design abstraction in a group modelling space may be better addressed in terms of the following:

Representation of multiple design viewpoints. Viewpoints of design are naturally brought in by members of a multidisciplinary team who work on various aspects of a building project. To enable contributions from all parties of expertise, these viewpoints must be represented in a group modelling space. This can be achieved by providing, firstly, design participants with an authoring tool to specify, and general *types of model constructs* that they consider pertinent to the modelling of a common generic structure. The idea is that different viewpoints can be associated with certain types of model constructs in a group modelling space. Instances of model constructs can be simulated to interact with form-giving forces or constraints applied and exhibit certain behaviours of deformation.

Secondly, to enable manipulation of the common structure from multiple viewpoints, participants need also to specify and generate *types of model connectors*, the devices to connect or disconnect instances of model constructs. Types of connectors will be used by participants to define and affect ways of manipulating parts of the common structures for various reasons. Note that model connectors are *domain-neutral* objects in a sense that they do not correspond to any specific building elements to be constructed in the real world. The same authoring tool can be used here, and the differentiation between model constructs and connectors is only meaningful to the human mind. From a computer modelling system point of view, model constructs and connectors are objects that have attributes and behaviours.

Representation of form-shaping forces. The flexibility or deformability is an important property of common generic structures, which is linked with group interaction in collaborative design. In a computer-based environment, form-shaping forces can be simulated in constraint-based graphic modelling systems. The provision of constraint systems in a group modelling space can be of two different kinds: *General constraint systems* supporting physical (or, more broadly, environmental) laws such as gravity, thermal energy, or acoustics, and *specific constraint systems* supporting intentional laws such as particular systems of spatial or shape grammars that are derived from bodies of architectural theories. The implementation of a constraint-based graphic modelling environment demands highly technical computing expertise; building designers are therefore not expected to build up, computationally, a constraint system by themselves. However, to illustrate the relevance, I shall introduce some exemplar constraint-based object-oriented graphic systems in the next section. In practice, it would be more appropriate for computing system engineers to develop computational models that are capable of interacting with instances of model constructs and connectors deployed by participants. We can think of systems of form-giving forces being implemented as visual computing modules that can be employed by building design teams as "plug-ins" in the group modelling space.

Common generic structures are pictorial and general. The preparations of model constructs, connectors and constraint systems are all geared up to the modelling of a common generic structure. Representation of the structures requires to be graphical and, at the same time, general to serve two purposes: In serving *all* members of a design team as a common design framework, representation of the framework is essentially *pictorial*, or, at least, *diagrammatic*. This ensures that all participants of different backgrounds can feel relatively *intuitive* in understanding their common design objectives as being visualised. Common generic structures are also essentially general in order to be refined or substantiated to different levels of design specificity. Therefore, its representation requires an order of generality to support the following flow of information:

	instantiation-of		*substantiation-with*	
CGS		**DS**		**DE**
	→		→	

Support for representation in individual modelling space

In teamwork, we should view that a participant's development of domain design solutions is not less important than that of common structures. To carry out more technical or specialised modelling tasks, members of a project team need to work with *individual* or *personal* workspaces that are not necessarily known and accessible to other members of the project team. A similar question with respect to support individual modelling is how to provide designers with an interactive authoring environment so that she or he can generate domain design constructs appropriate with regard to the design tasks at hand. This requirement gives rise to the following sub-issues.

Representation of individual object worlds. This includes, firstly, a set of personal design constructs for generating and manipulating domain design expressions, and secondly, domain-oriented constraint systems deployable for form-finding in domain design developments.

Generating domain design expressions with reference to derivative structures. Designers are allowed to work out domain designs in relation to the derivative structures of their design interests. Similar referencing mechanisms are already available in major CAD packages such as AutoCAD or MicroStation (more details on this in Chapter 7). Used as a visual interface, the working area for editing domain designs can be *overlaid* or *juxtaposed* to the areas that hold instances of derivative structures as design referents.

Generating domain design expressions by substantiating derivative structures. Instead of referencing, this is a designer making direct use of derivative structures imported from a group modelling space. Basically, this involves designers enriching, refining, or transforming derivative structures into domain design expressions filled with domain-specific design details. In computing terms, the process of design substantiation can be simulated in the computer programming concepts of *parameterisation* and *instantiation* as developed by Joseph Goguen and Timothy Winkler in their implementation of the programming language OBJ3 for parameterised programming (Goguen and Winkler 1988). Rather recently, adopting the MicroStation/J platform and an end-user visual programming methodology, Robert Aish has developed the *CustomObject* framework in which designers can create their own novel modelling functions not provided by the CAD system originally (Aish 2000).

Support for coordination and negotiation

The facilities to enable abstraction in a group as well as multiple individual modelling spaces discussed above can be seen as the *infrastructural* supports for project participants setting up group and individual workspaces. In addition to the *infrastructural* aspects, there are other system requirements of more dynamic features to be considered. The third issue is concerned with a system's ability to support *coordination* and *negotiation*, which are described in details in the following components.

Detection of change in common generic structure. As long as its deformability is maintained in the group modelling space, the state of a common generic structure is always subject to participants' manipulations from different viewpoints. It is the evolving of a flexible generic structure that gives rise to the dynamism of teamwork in the top-down scenario. As discussed earlier in the constraints of coordination and negotiation, this is probably one of the most challenging requirements for developing computer support with which the design team can better manage the group dynamism in their collaborative working. For a design environment to serve in the dynamic situation, it will be concerned firstly with gathering the facts about changes in the state of a common generic structure.

We may define a state of generic structure as a two- or three-dimensional configuration of instances of model constructs and connectors under the influence of a system of form-giving forces activated in a group modelling space. A state change in a common generic structure can therefore be defined as a change in an existing configuration due to a net effect of some participant's or participants' model manipulations under the form-shaping forces applied. A design support system's ability to keep track of state changes in common generic structure lies in whether the system can generate information regarding the differences in the configurations resultant during a session of group modelling. The change detected is essential for the system to trigger further mechanisms for communications such as maintaining derivative relations and delivering messages to participants for their maintaining referencing relations.

Maintaining the relation R_d in R_d[CGS, DS]. In developing domain design solutions, participants import derivative structures from a state of common generic structure as design guidelines or references. Since the state of the shared structure may keep changing, it will be a useful support to the project participants if a computing system can react to inform group members in a timely way regarding the changing states of the derivative structures under their uses due to the state change in common generic structure. This requires a system to keep a record of the relations R_d between derivative and common structures and compute updated states of the derivative whenever the common gets changed. Apart from the states of ***CGS*** and ***DS***, two other computer representations (or, registrations) are required for a system to maintain the relation R_d.

In acquiring derivative structures, designers perform certain *spatial operations*, such as *projecting, subdividing*, or *slicing*, upon shared generic structure. A representation of derivative operation thus consists of the performers and the spatial operations performed, i.e., the identities of the persons (or, the individual modelling spaces) and the derivative operations as made known to the system. Furthermore, the information about the time and location relative to a state of common generic structure in which a derivative operation takes place is also relevant in maintaining the derivative relations specified. This will constitute a registration of the temporal spatial location of deriving within the system.

Message sending for maintaining the relation R_f *in* R_f*[**DS, DDE**].* The development of domain design expressions can be perceived and interpreted by project members working in particular project domains as *design consequences* in respect of a state of common generic structure. Through reviewing the resultant design development, some members may well be motivated to make changes in their current domain design expressions. As shown earlier, there exists the referencing relation, R_f, between derivative structure and domain design expression. Given a change in domain design intended by its designers from a particular design perspective, a referencing relation may not be maintained if state changes in the underlying derivative framework and, hence, in the common generic structure, are not reflected correspondingly.

To assist designers' management of the consistency in local design developments, individual modelling spaces should be equipped with a system capability of not only allowing for designers to freely make changes in domain design expressions, but also assisting the individuals in maintaining the referencing relations set up earlier. Two kinds of system functionality can be instrumental to achieving such a capability. A detection mechanism for determining state change in derivative structure can be useful. This is similar to that of detecting state changes in common generic structure, but the detection functions need to be installed locally as individual modelling spaces may be distributed over a number of separate working sites. Secondly, a message-sending mechanism is required. When a state change in derivative structure is detected, a message is generated and sent to the group modelling space for activating corresponding state change in the common generic structure.

Suppose that a message sent from an individual modelling space is received and processed in the group modelling space, a change in the common generic structure will be generated by the system. Due to the mechanism of maintaining the derivative relations declared earlier, further messages containing the information regarding *pending* changes in derivative structures will be generated and sent from the group modelling space to relevant individual modelling spaces. In this way, participants involved in the change shall be informed. The detection and message delivery mechanisms described here seems to suggest that a *local management agent* may be set up in an individual modelling space which is the sole information space for the agent to serve.

Communication channels for resolving design conflicts. If the coordination constraint, as described previously in Section 6.1, cannot be satisfied, negotiation among the individuals involved in the disagreement is required to resolve the conflict. Since how the conflict may be resolved is highly non-deterministic, a system is not expected to automatically detect any conflicts occurring and resolve them. In practice, this may have to be left to the human participants to decide if coordinating or negotiating is required. If coordination is to follow, there is no need for participants to express further individual assessments of the (changed) state of common structure, and corresponding changes in domain designs should be carried out in individual modelling spaces respectively.

More problematically, if negotiation is to take place, participants involved will need to discuss disagreement with one another.[35] This demands a design environment to provide participants with communication channels with which they can discuss the conflicts directly or indirectly and resolve the differences in recognising the state of common generic structure until the shareability is re-established among members of a design team.

6.3 Constraint-based graphics and other related research

As we can see in Chapter 2, the recent technological trend in developing network applications in support of collaborative design seems to take on two main approaches: (1) developing virtual shared workspaces for real-time group drawing; (2) implementing a document or work flow management model within a conventional CAD system. The question is if the current approaches suffice to support innovative multidisciplinary collaborative building design that requires spontaneous interaction between design integration and design distribution. To better facilitate the spontaneity and complexity of group dynamism required by longer-term multidisciplinary collaborative design, an alternative approach is needed. It seems to me that we need to look into system development capable of supporting a group modelling space in which no predefined group hierarchy or protocol is imposed and multiple viewpoints as acted by team members are supported. I also consider that the issue regarding what relations can be established between a common framework made of simple generic constructs and multiple detailed expressions drawn on distributed domain-specific design knowledge is largely unexplored in a design computing environment. Before the proposed system architecture is presented, let us look at related research in three areas.

Based on analyses of organisational problem-solving in scientific communities, Susan Leigh Star describes the concept of *boundary objects* and suggests that the concept can be further developed into an experimental data structure for distributed artificial intelligence systems (Star 1989). Star identified types of boundary objects that served the scientific communities in solving heterogeneous problems. Notably, the boundary objects bear similar properties to those of common generic structures in collaborative design as seen above. Star describes boundary objects as follows:

Boundary objects are objects that are both plastic enough to adapt to local needs and constraints of several parties employing them, yet robust enough to maintain a common identity across sites. They are weakly structured in common use, and become strongly structured in individual-site use."

It seems reasonable to suggest that the flexible generic structures in collaborative building design can be considered as another type of boundary objects but with a *generic-*

specific adaptation instead of a weak-strong one. Although the general properties of boundary objects were identified, no computational representations of these objects have been implemented as alternative data structures for distributed AI applications.

In relation to the issue of modelling flexible generic structures in computers, research on advanced computer graphic modelling has shown us some technical possibilities of visualising objects that are able to respond to applied forces or constraints in a natural way. Three research results are particularly relevant to the deformabilty or flexibility in common generic structure. Terzopoulos and others have employed the theory of elasticity and produced elastically deformable computer models (Terzopoulos and others 1987); Witkin and others explored the representation of geometrical constraints as energy functions that behave like forces pulling and deforming parts of an object model into place (Witkin, Fleischer, and Barr 1987). Platt and Barr explored force-based constraint methods that can build up several physical properties into flexible graphic models (Platt and Barr 1988). It remains to be seen how the kinds of computational graphic modelling techniques as developed by these researchers may be applied to the context of collaborative building design, especially, in accommodating the requirements of generic and multi-perspectiveness. As for setting up domain-specific constraint systems in distributed individual modelling spaces, some examples of how design constraints can be represented on a smaller computation system but dedicated to certain design topics were developed by Mark Gross and others (Gross et al. 1988).

I have mentioned in several places the relevance of constraint-based computer graphics to the modelling of common generic structures. Research and development on object-oriented and constraint-based programming has been very active in recent years and it may continue to attract more research in the years to come. From a design computing point of view, it seems to me appropriate to adopt constraint-based object-oriented graphics as a core strategy of representing or simulating a system of form-giving forces that becomes the backbone of an interactive graphic modelling environment. However, I should say that combined object-oriented and constraint-based programming is certainly not the standard way of developing a system in accordance with the top-down approach but a suitable one that may lead to further development of a more complete system. In what follows, we shall look at an exemplar system that will illustrate some of the key features in object-oriented constraint-based graphic modelling.

The *Briar* drawing programme was developed as a research prototype by Gleicher and Witkin at the Carnegie Mellon University (Gleicher and Witkin 1994). How to lift the barriers of creating constraints, solving them, and presenting them to users were some of the main issues addressed in the development of *Briar*. An important system requirement studied by the researchers was that "not only must a system be able to solve constraint satisfaction problems, but it must make it easy for users to specify, debug, and edit constrained models."

Figure 6.1 shows a snapshot of the user interface of *Briar*. The user is presented with a palette of drawing objects and functions with which drawings with their underlying constrained models can be created and edited. To have an overall picture of what can be done with the drawing programme, some of the main systems features and functionality of *Briar* are summarised as follows:

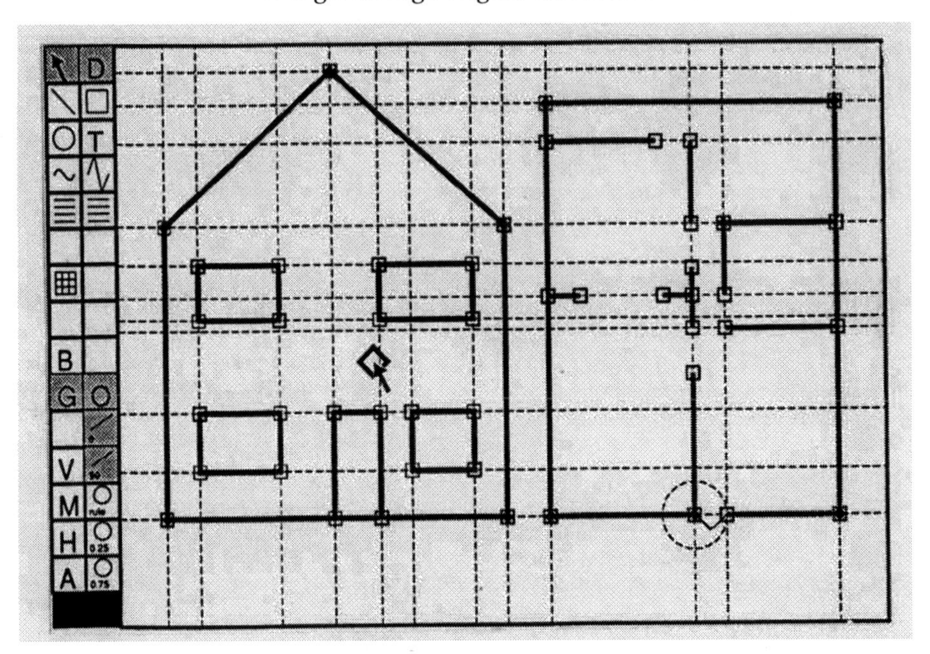

Figure 6.1 *The user interface of the Briar drawing space.*

Drawing primitives. Like most object-oriented drawing editors, *Briar* has a limited set of drawing primitives: line segment, rectangle, circle, and text. For each instance of the drawing primitives, there is a number of pointers associated with the image of the object. A rectangle drawn in Briar, for instance, has four pointers any of which can be clicked and dragged and results in a rectangle of different size or location.

Constraints editing. The basic mechanisms used by *Briar* for users to specify and edit constraints in drawings are snapping and dragging (or Snap-Dragging in short). A snap-dragging operation is enacted when pointers of a drawn object being dragged sufficiently close to a vertex (i.e., an intersecting element, end point of line segment, etc.), edge, or curve in the scene, the cursor snaps to it. Snap-dragging provides a basis for graphical representation of constraints. There are two basic constraints maintained by Briar: point-on-object and point-to-point.

Augmented snapping. For every snapping operation, there is a displaying of a (graphical) symbol indicating the constraints, which the snapping implies. The visual cues and feedback mechanisms provided here include the following:

- A symbol of an empty square is shown when a point is snapped to an object (a drawing object or an alignment object);
- A symbol of a filled diamond is shown when a point is snapped to another point (a point can be an endpoint of a line or a point in a circle, for example);

- A symbol of a dashed circle with an indicator showing its radius is shown when the constraint of distance-control is applied between two pointers;
- The cursor changes shape depending on whether it is snapped to curve (shown as a cross) or edge (shown as a diamond with a cross in it);
- The object snapped to is brightly lit;
- If several objects are close together (cluttered), the one desired can be selected by cycling, which can help to solve the problem of ambiguity.

Constraint solving techniques. The constraint solver of *Briar* is only intended to maintain constraints, rather than to establish them. That is, the constraint methods are used to move the drawing through configurations, which are consistent with the constraints, (i.e., the relationships already established by users using snap-dragging) rather than moving from configurations inconsistent with the constraints to ones that are. This position allows the use of differential constraint method and avoids the difficult problem of solving non-linear equations from arbitrary starting points. The differential methods operate by controlling how the objects move by specifying the time derivatives of their configurations. The input to the methods is typically a desired motion (e.g., a point tracking the mouse, and a set of constraints whose values are to be maintained). The methods then solve for the time derivatives of the configuration that achieve these. Linear equations that specify the time derivatives of the constraints are maintained to the effect that the non-linear equations on the configuration are enforced.

Constrained dragging. In *Briar* there is a category of constraints characterised as "lightweight" constraints. Dragging and tacking are two examples of such constraints. The short-lived dragging constraint is achieved by temporarily constraining the point being dragged to follow the mouse. Tacks hold a particular point in place. Figure 6.2 shows a sequence of interactions in which the user is playing with a constrained model of engine through constrained dragging.

6.4 GAUDI: A proposed system framework

In order to inform a more specific research programme of developing computer support for aspects of group interaction in the top-down scenario, I shall in this section propose a system framework. Because of the connection with the case study of the Colonia Güell Church project presented earlier, the proposed system framework is referred to as *GAUDI* in the following discussion. The goal is to develop an intranet-based pilot CAD environment for supporting group modelling of shared design frameworks in connection with distributed modelling of domain-specific detail designs. Given that an object-oriented approach comes close to the designer's need for modelling design with structured representations, and that constraint-based methods can deliver the kind of responsiveness or flexibility to the group processes of modelling common generic structures, it seems to me logical to set the course for developing the *GAUDI* framework along the line of an object-oriented constraint-based graphics modelling and mediating environment. By achieving *GAUDI*, we can be in a better position to investigate in greater details regarding how end-user induced teamwork interaction can be maintained

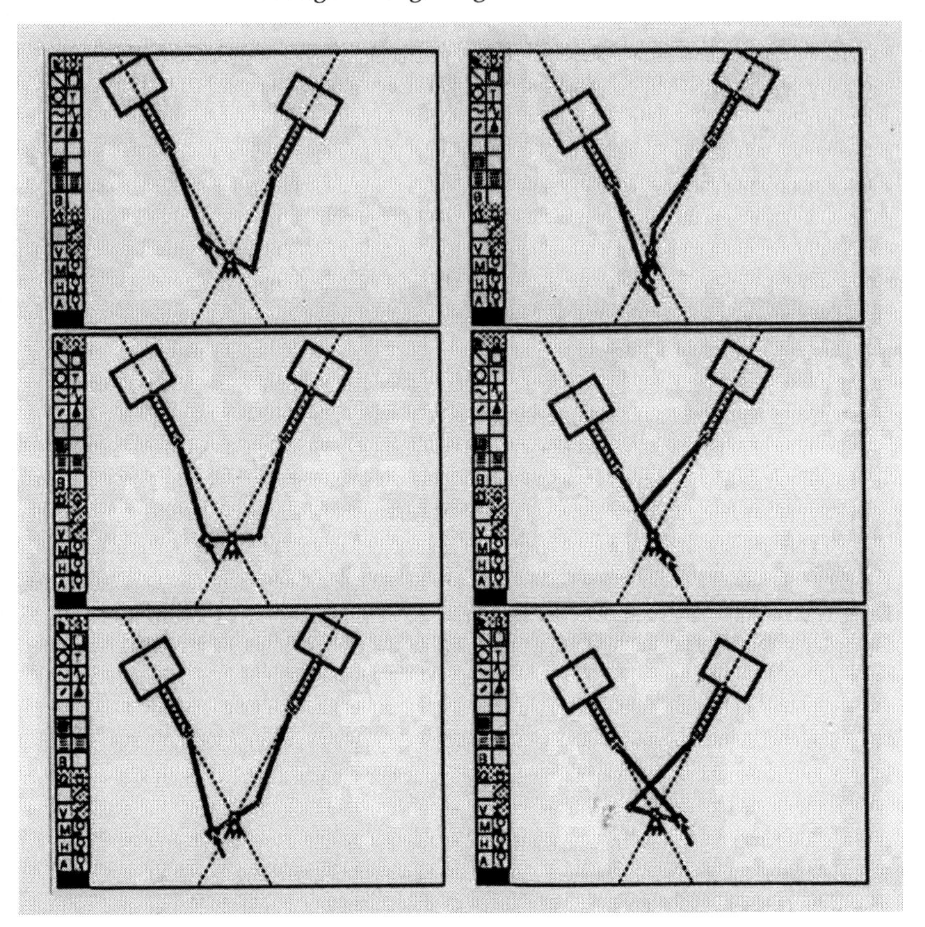

***Figure 6.2** Interacting with a constrained model of engine in Briar.*

efficiently and sustained within a collaborative design environment over a project's lifetime.

In spelling out what system components and architecture the *GAUDI* framework would consist of, I shall begin with the aspect of supporting design representation, then the aspect of supporting design communication that centres around the design representations supposedly created by the end-users (i.e., the designers). More specifically, I shall describe the *GAUDI* framework in three related parts: (a) representation of multiple viewpoints; (b) construction of a (funicular) constraints-maintaining system; and (c) design of cooperation supporting mechanisms.

a Resources for representing multiple viewpoints

We first describe a strategy of establishing a group modelling space in which flexible generic frameworks can be constructed collaboratively by team members. The group modelling space will provide resources for introducing and representing multiple viewpoints as enacted by various individuals. Following the group modelling of the funicular framework as an example, a strategy of achieving this is to provide two kinds of resources (Figure 6.3):

Model construct specifier. Model constructs are the basic elements for the design team to build up their own common generic frameworks. Basically, they are graphical objects in the sense of object-oriented graphics. Instances of model constructs have attributes that determine their shapes and methods that define ways of construction and manipulation. A Model Construct Specifier, a kind of end-user authoring tool, provides a set of system-defined parameterised graphics primitives which can be instantiated (interpreted) by end-users into more specific model constructs that, in their views, are pertinent to the project at hand.

Repository of connectors. A repository of connecting devices will be employed by users in making connections among different types of model constructs to the effect that a constructed common generic structure can be (directly) manipulated for more than one reason. Notably, model connectors are not parts of the generic structures per se, but exit only to be connected to. Unlike model constructs, model connectors are neutral objects in the sense that they are not deployed to represent any particular design viewpoints.

b Constraint-based graphic modelling

Assume that the members of a design team work with the gravitational forces on Earth which govern the shape and behaviour of funicular structures. This can be computationally implemented as a constraint-based modelling environment. Mathematics equations of funicular structures have been established which are useful references in the design and implementation of a constraint solver. A constrained model of funicular behaviour can be interacted with to establish an equilibrium state for every funicular structure constructed. Figure 8 shows a general funicular structure of a uniformly loaded cable with both supports on the same level. Following the equation established by Daniel Schodek (Schodek 1980), in a constraint-based approach to the modelling of a simple cable-like funicular structure, the set of objects involved would include:

A, *B*: left and right supporting ends with coordinates
L_h: the distance between A and B
h: the maximum sag
L_{Total}: the total length of the cable

The relations among the above set of objects are governed by:

$$L_{Total} = L_h(1 + \frac{8}{3}h^2/L_h^{\,2} - \frac{32}{5}h^4/L_h^{\,4})$$

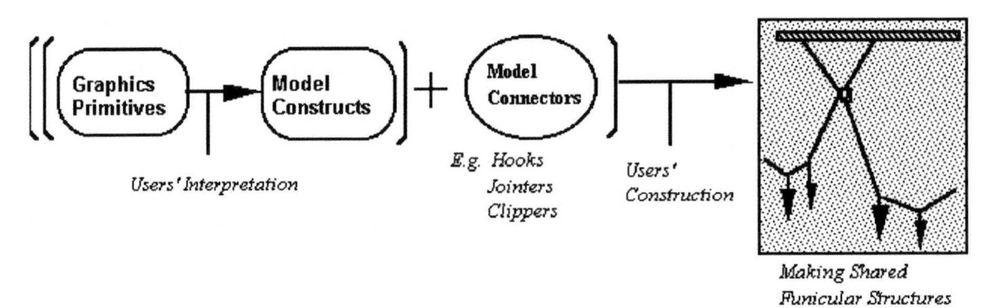

Figure 6.3 *The proposed resources for representing multi-viewpoint in GAUDI.*

Figure 6.4 shows a general funicular structure of a uniformly loaded cable with both supports on the same level.

The above is simply to illustrate an example of implementing a constraint-based graphic system for group modelling of flexible generic frameworks. There can be other kinds of form-shaping forces at work, and it is certainly not reasonable to assume that building designers are able to implement such constraint maintenance systems. As a general principle, we suggest to treat constraint solvers as particular system modules that can be plugged into a group modelling space when delivered by computing system specialists.

c Coordination and negotiation supporting agents

Given the resources available for representing multi-viewpoint and constraint-based graphic modelling, group dynamism in multidisciplinary collaborative design may naturally arise. Regarding this, we propose to experiment with the following communication supporting mechanisms as computer-based agents for mediating team coordination and negotiation:

Derivatives register: An agent for users to register how derived structures are deduced from a common generic structure. A scheme of registration can take the form: which can be read formally as "a registration *R* regarding the deriving of DS owned by some designer with the identity *id* is equal to the application of the derivative operation onto ***CGS*** with a state *s* under the perspective of the designer with the identity id."

Derivatives updater: An agent responsible for computing registered derived structures into their updated states at some moment. The agent is activated by receiving message indicating a change in the state of CGS has been proposed by some individual.

Task messenger: By cooperating with the derivatives updater, the project messenger checks if an individual modelling space is involved in proposed design change. Messages containing the information about updates of derived structures in use, the identity of the individual who initiated the change, and a request for giving judgement will be sent to the addresses of those individuals involved in the proposed change.

Conferencing manager: Design judgements are collected and interpreted by a conference manager. If an agreement is reached, coordination follows; otherwise,

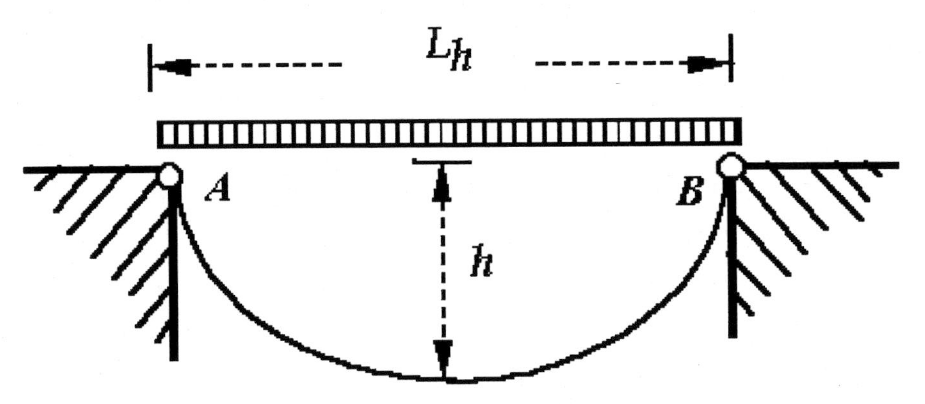

Figure 6.4 *A diagram for a simple cable-like funicular structure (After Schodek 1980).*

communication channels (like shared sketchpad, text editor, chat box, etc.) will be opened up for negotiation to follow.

A system architecture for organising the above system components is set out in Figure 6.5. The pilot system is essentially a distributed client/server architecture with all the representation resources and the communication-supporting agents implemented in the server station. Among a cluster of client stations, multiple individual working platforms can be connected to the server station via an intranet-based communications network. Each working platform can be further linked up with a local knowledge or data base used for a particular design domain. User interfaces supporting various modes of group interaction (i.e., whole group, sub-group, one-to-one, and private) will have to be developed under a single simple scheme equally presented to every system user as a participant of a building design team.

On the basis of the *GAUDI* framework and the system components proposed above, I expect that an implementation of the pilot collaborative design modelling environment will open up a number of research topics for further investigation into *users-centred* collaborative design computing:

Design of multi-user interface. A multidisciplinary design team may employ a collaborative CAD system for both synchronous and asynchronous collaboration. In this user context, the usability of a system is critically related to the design of multi-user interfaces in which all networked workspaces should be provided with an interface consisting of three compartments: (a) Group modelling space for facilitating "What I see Is What You See;" (b) Individual modelling space for supporting domain-oriented private design working which is not necessarily shared among team members; and (c) Communication channels for enabling informal one-to-one or one-to-many remote design discussion. No predefined team structures should be built into the interface such that each participant can make contribution to teamwork on equal terms.

Distributed uses of common generic frameworks. This aspect of further research is to explore what computational links may be established between a common generic

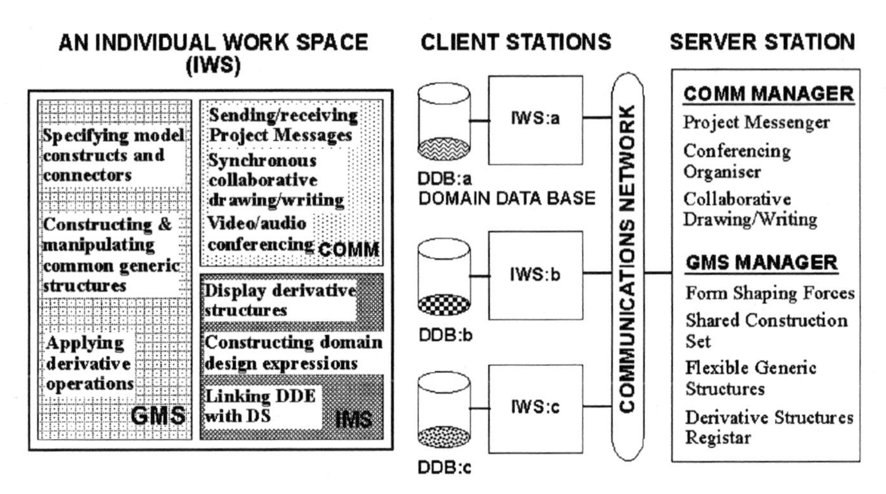

Figure 6.5 A proposed system framework GAUDI for multidiciplinay collaborative building design.

framework modelled in the group modelling space and domain-oriented design expressions generated in multiple individual modelling spaces. Currently we think that the links can be made on the basis of two operations: (a) parts of a common generic structure are substantiated into domain design expressions with domain design constructs maintained in local design knowledge/media bases; or, (b) domain design expressions are constructed with reference to some parts of a common generic structure. The main task is how to define the operations of substantiating and referencing in computational terms. These operations are also related to the functionality of the mediating agents described earlier.

Interaction between design integration and design distribution. Multidisciplinary synthesis of built form is an open process associated with group work on the development of wholes (i.e., integration) as well as of parts (i.e., distribution or articulation). In this chapter, I propose an approach that does not prescribe a fixed model or relation between wholes and parts to be mapped onto either a hierarchy of teamwork or the data structure of a product; system development ought to address the requirement that newer wholes may emerge constantly from continuous distributed developments of evolving parts, and vice versa. In my view, the challenge of development of *responsive teamware* for innovative multidisciplinary design of complex artifacts, like buildings, lies in the support for the spontaneity of the interaction between integration and distribution.

6.5 Operations, images, and collaboration

Through Chapters 3 to 6, we have looked at case histories of teamwork in design exclusively upon the design expressions generated among professional designers. Collaborative design to many people may include a far wider scope of participation including, for example, clients, end users, or governmental representatives, etc. How and what to support teamwork involving both professional and non-professional designers is certainly an important and challenging subject area, and it will surely demand many more research efforts to specify the representation and communication requirements in a clear and systematic manner. Given my limited research resources for working on the project, a research into a wider scope of participation was not attempted. Instead, I have focussed on the various examples of diagrams, drawings, and models produced by professional designers for real world building projects, identifying the factors of communication and coordination in collaborative design. Drawing on the source materials collected, I have tried to make explicit how the design developments are related to the interaction among participating design worlds. Like many other collective human activities involving smaller groups, I think we can talk about *group dynamics* in collaborative design. And it is important to accommodate the group dynamics found if we are to succeed in developing a team-oriented computer aided design environment to support design collaboration.

However, it should be clear now that the group dynamics researched for collaborative design has a different emphasis from that of social psychological enquiry, which is concerned more with the knowledge about social psychological forces associated with groups (Cartwright and A. 1968, 4). In creative collaborative design, it more appropriate to look at the *information* factors associated with team working. In particular, identifying the patterns of information flow among the various factors can contribute to descriptions of how designers of multidiscipline work as a team in achieving common project goals. In unravelling general factors of group dynamics in collaborative design, I introduced the basic notions of *modelling spaces* and *modelling acts*, and a combination of the modelling spaces and acts leads to further notions of *design information carriers* as characterised in the various design states. It is with these abstractions that I arrived at an outline of the key aspects of group dynamics in collaborative design.

By organising these information factors together, two patterns of team working in collaborative design are described in the current studies. Due to the different features of information flow observed from the case histories, the two patterns of collaborative design were characterised as *top-down* and *bottom-up*. As mentioned before, these patterns are possibly two among many others if the current scope of design studies is extended to examine more cases. More importantly, for the purposes of conducting system research, it is the sets of requirements for collaboration that we should pay attention to rather than the named patterns as distinctive teamwork styles.

Collaborative design in the bottom-up scenario starts with participants introducing domain primitives and operations with which domain expressions are modelled in individual workspaces. Collaboration on achieving integrated conceptual structures or schemata is initiated by project participants presenting domain design proposals in a common workspace. Using the shared design constructs or primitives and operations, domain design expressions can be combined and integrated into joint project work as a whole. On the basis of combined individual design contributions, common images may be discovered at some stage by some members of the design team and subsequently are shared by the rest of the team. Common images emerged from specific project contexts can serve members of the project team as shared metaphors that provide powerful conceptual frameworks for articulating the relationships between parts and wholes.

In comparison, collaborative design in the top-down scenario starts with group modelling of flexible generic structures at the inception stage of project development. The knowledge and skills of constructing such structures may have been cultivated by the design team over many years of practice that form a particular philosophical or technological standpoint which the group is strongly associated with. The properties of being generic, flexible (deformable) and multi-viewpoint are crucial for the common structures to sustain group dynamics in project development. Team members employ spatial operations allowed by certain systems of form-governing forces or rules to manipulate the state of common generic structures. By extracting derivative images from parts of the common structures according to project participants' design interests/responsibilities, domain design expressions can be further developed and elaborated in distributed individual modelling spaces. Designers then receive consequences of making domain design changes by querying states of common images, which may activate interpersonal coordination.

If we take the flow of information as a measurement, the top-down oriented collaboration is initiated by a beginning whole at the top that leads to distributed developments of parts at later stages; the bottom-up oriented collaboration, on the other hand, is motivated by participants seeing the connections among individual design parts generated from the outset that leads to integrated wholes in the end. Parts and wholes in both scenarios are subject to continuous changes involving communications and interactions among team members.

Though exhibiting different patterns of information flow, top-down and bottom-up collaborations demonstrate similar preconditions for teamwork to take place. We see that specific or concrete goals of collaboration (e.g., definitive descriptions or depictions of final design products) are unknown to or cannot be clearly defined by any of the project participants at the outset. Heterogeneous systems of representation and action employed by individual members are necessarily involved. The heterogeneity appearing among participating design worlds is what we observe externally; however, it should be pointed out that there is also (hidden) homogeneous similarity among individual design worlds which makes understanding across heterogeneous differences possible. Not to mention that the co-existence of the two ends is certainly a simplified view of reality. We also observe that no predefined scheduling schemes are applicable to determine how participating design disciplines should coordinate with each other during collaborative sessions, i.e., the strategies for specifying and satisfying precedence constraints of fulfilling cooperative processes cannot be determined by particular individuals in advance.

Working under the above preconditions, designers often commit themselves to producing sorts of design information in the forms of sketches, drawings, physical models, specifications, etc. These are the outcomes of individual as well as group working. More specifically, the production of two kinds of information can be generally recognised as the goals of teamwork. Firstly, the information conveys shared conceptions of *unity* in design artefacts yet to be realised via construction in the real world.

Presumably, design expressions of unity can be reached collectively by participants on the basis of, perhaps, their common life experiences and knowledge in dealing with general issues, such as form, movement, human biological or ecological needs, and so on. Secondly, the information contains separate records of *specifications* addressing requirements occurring in particular design domains. The design specifications are produced on the basis of separated individual expertise in dealing with technical and detail designs, such as structural, lighting, mechanical services, and so forth.

My current design studies presented in this book seem to suggest that the two goals of collaborative design can be parallel to each other in the sense that no one goal is fully determined by the other. At best, we can say that they can be influenced by each other in the course of project development. This is important because the parallel demands more dynamic approaches to addressing how the working relations among team members may be computer supported. The differentiation between top-down and bottom-up patterns is based on observing case histories of building design, showing the different representation and communication requirements of sustaining design collaboration. This should not lead to the development of a computer-based environment that serves solely either the top-down or the bottom-up approach. Design collaboration and computer system development in the real world is more complex than this. A design practice may adopt different strategies for different building projects from time to time, and, therefore, they may demand different system capabilities of supporting team design in varying project contexts. Is it possible to accommodate both the top-down and bottom-up requirements within a single system development? Perhaps, we can have better ideas by building up a collaborative design modelling environment by configuring a range of system components that are designed to address certain parts requirements across the different patterns.

Summary

Following the top-down pattern described in Chapter 5, this chapter develops an account of collaborative design activity, which is featured by group members' joint substantiation of common generic structures. To show that the funicular structure seen earlier is not the only case, we present further evidence of generic frameworks from a collection of taxis schemata. Structural objects of this nature play an interesting role in collaborative design.

They can be created, manipulated, and evolved by all members of a design team from different design viewpoints. Once constructed in a shared workspace, common generic structures can function as communication frameworks upon which participants interact with each other in the course of developing domain-specific designs.

To unravel what conditions the top-down approach, we carry out a situation-theoretical analysis of an abstract scenario. A scheme of information flow among the situation types classified indicates two transitions where group interaction might have to occur. One is the generation of a generic structure from a shared construction set, where participants need to jointly experiment with *model connectives* so that instances of various model constructs can be linked in a way responsive to the action of form-giving constraints. Much more dynamic group interaction can occur when any of the group members attempts to make design changes. Motivated by domain design judgements, one or more participants may go on changing parts of a structure which may cause further changes to follow in other members' work. The continuation of teamwork is therefore constrained by *coordination* or *negotiation*.

We derive three issues to be addressed when computer-based group design environments are to be developed. In the representation aspect, general schemes can be developed to support the construction of the various components of individual as well as group modelling spaces. In the communication aspect, mechanisms such as *state change detection* and *relation-maintaining message delivery* can be looked at to facilitate design coordination and negotiation. A potential technological basis for developing collaborative design computing of this nature can be found by interrelating some of the related research done in computer graphics, distributed artificial intelligence, and collaborative drawing support tools.

7 Current Industrial CAD Platforms for Team Working

The multi-user collaborative drawings tools seen in Chapter 2 were mostly developed by researchers working in academic settings. In this chapter, we shall look at the teamwork supporting features currently offered by some of the most popular industrial CAD packages. The early development of the research prototypes has been mainly driven by the academic researchers' interests in advancing technological innovations that would demonstrate the potentials of computer supported cooperative working. In contrast, the development of teamwork supporting facilities in the CAD software industry presents a very different scene. The earlier generations of commercial CAD systems were developed and deployed predominantly as individual computing platforms; and each of the CAD packages has been built and evolved from a distinctive proprietary digital data or file format (e.g., the DWG format of AutoCAD, to say the most known).

If a CAD package survives the market forces long enough and acquires a certain mass of end-user group, the unique CAD file format becomes inseparable from the commercial brand established. From a commercial point of view, it may be necessary to abide by the business logic of maintaining the file format technology and any basic software design associated with it. However, the legacy of a file format can induce inertia or resistance upon radical system innovations that may render the existing software version incompatible or too expensive for current users to have their software versions upgraded. Therefore, it is not unusual to see that most newer versions or upgrades of CAD packages are minor 'tweaks' rather than real innovations. Commercial CAD vendors are also concerned with the issue of leaving a steep learning curve with users because of radical new system features. Given that mastering a modern sophisticated CAD package is by no means an easy task in contemporary building design practice, CAD managers/users are conscious about the economy of investing in new system functionalities that may not return significant benefits. There is also the aspect of third-party developers who produce software marketable in connection with the host platform. Unlike pursuing system developments in an academic setting with almost free hands on innovation, commercial CAD software developers and vendors are most concerned with how to maintain a certain level of continuum with the existing system facilities when new features are introduced.

Most current commercial CAD platforms have their origins as single-user systems as they first appeared. Not until recently, teamwork-enabling functions in CAD have seldom been considered as major marketable features by the vendors. Mainly due to the recent global spread of the Internet and the worldwide enthusiastic acceptance of the Web, there appears a discernible trend in the CAD software industry to deliver collaborative design capabilities in response to today's new ways of working and communications enabled by digital networks.

Intended as a complement to the previous survey on the early experiments of group drawing tools, this chapter presents a review of the teamwork-supporting facilities currently offered by three popular commercial CAD platforms: ArchiCAD, AutoCAD and MiroStation ProjectBank. There are other CAD packages also widely used by design practitioners and students such as form•Z of auto•des•sys, Inc. and VectorWorks of Nemetschek N. A. Inc. (formerly, MiniCAD of Diehl Graphsoft Inc.). However, at present, these CAD packages do not provide special facilities for supporting collaborative design. In addition to reviewing the specific industrial CAD platforms for teamwork, we shall look at the recent industrial initiatives of developing open-access data schemas for future pan-industry information sharing and transactions over the Internet. The emerging initiatives are based on adopting the eXtensible Markup Language (XML) as the common machine as well as human readable language, which is used to describe semantic content of information generated in the domain of building design, construction, and operation. We shall look at three XML-based schemas in some detail: aecXML, bcXML and gbXML, and discuss its implications for future development of commercial software applications aimed at the building industry.

7.1 ArchiCAD for TeamWork

The basic concept underlying the ArchiCAD for TeamWork (version 5.1) is to support collaborative working through high-level file sharing over a (non-server based) computer networking (Microsoft's Windows or Macintosh OS). With this CAD platform, team working is initiated by a "Team Leader" who creates a central master project file and an associated administration file on any computer connected to the file-sharing network. Other designers can then get access to the central master file and sign in as "Team Members" to "reserve" particular working areas. A member's singing in and reserving an area on the central project file will result in an individual "Workspace" designated by a combination of "Stories, Layers, Marquee Polygons," or others (see Figure 7.1). Any overlapping of reserved work areas will be detected by the software and is not permitted. Working either online or offline, a team member can see the entire project file while working in his or her own individual workspace, but changes can only be made onto the elements designated within the individual workspace. Team members' contributions can be gathered at any time when members initiate the "Send/Get Changes" command, which effectively update the existing master project file with elements newly created or modified by each individual workspace. Since no two reserved working areas overlap, inconsistency among individual design contributions is not an issue of concern.

On the basis of the master file issued by a project's team leader, ArchiCAD for TeamWork assumes that setting up of multiple individual workspaces will be conducted by the design group in a collaborative manner through, mostly, face-to-face communication and negotiation. Alternatively, the team leader can adopt a top-down project management scheme, pre-assigning who will be doing what with

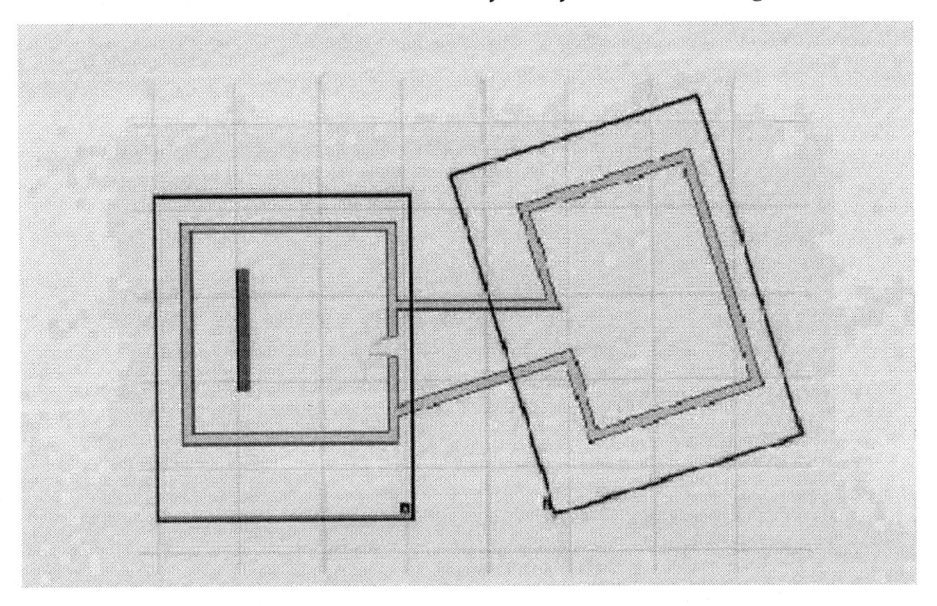

Figure 7.1 *Creating individual Workspaces in ArchiCAD for TeamWork by reserving an area of interest in relation to the central project file. In this case, two designers use the Marquee Polygon tool to reserve/declare individual working areas.*

regard to the master project file. However, an interesting feature of ArchiCAD for TeamWork is that it allows for a team member to "Sign Out" or "Release" his or her reserved working area or element so that other team members can sign in to continue working on that particular area or element. This is considered by the software developers a step forward in comparison with more conventional file referencing capability (Szovenyi-Lux 1997). Several examples of how to move around design elements between individual workspaces so that team members can co-work on the same elements in a "Give-and-Take" manner can be found in the ArchiCAD's user manual (see Chapter 5, Graphisoft 1997).

As pointed out candidly in the software's literature, to take full advantage of the collaborative tools provided by ArchiCAD for Teamwork, members of a design team should have face-to-face contact when working out details of project sharing so that collaboration in design can be carried out more effectively. It is not clear to what extent the effectiveness of the TeamWork package may suffer if the team members are geographically dispersed and cannot sustain direct communications of a high quality. One can imagine the inconvenience caused by not being able to receive an immediate response from other team members when initiating changes about reserved work areas or elements. A notifying or informing facility can be a valuable addition to a future release of the TeamWork platform. If two designers cannot inform each other face-to-face, activating message sending without leaving the TeamWork environment can come in handy, especially if the message editor can also import or attach the design parts and context in question. Further system development may consider the possibility of an *intelligent messenger* that can automatically deliver notification messages to the parties involved in the same project work. With the messengers in action, rapid inter-workspace design communication can still take place even if designers work apart geographically.

If a stable division of project sharing plus regular direct face-to-face contact among team members can be achieved, ArchiCAD for Teamwork can provide an elementary platform for the top-down scenario of collaborative design as discussed earlier. However, one major conceptual issue not addressed by the current release of ArchiCAD for TeamWork is that a clear-cut division of parts may not be always attainable or desirable in a building design project. Team members may need to work on areas or elements simultaneously that actually overlap with each other to allow for exploratory design development pursued by several individuals. Of course, a more relaxed approach to sharing parts of a project could lead to inconsistencies among individual design proposals. However, from a design viewpoint, most designers would consider these inconsistencies as being temporary or transitional, which may not necessarily lead to a disastrous situation. To take perhaps a more positive stand, instead of limiting the ways designers may co-work from the outset, a collaborative design environment can be built to detect such inconsistencies at the background and then inform the parties involved to resolve the inconsistent design states. Like most commercial packages, ArchiCAD for TeamWork seems to adopt a rather conservative view that a CAD system is a machine mainly for making production drawings. The difference here is that production drawings are now drawn by a team of designers who need to declare or specify explicitly an organisation of dividing up a project to ensure that the resultant group drawings are inconsistency-free.

7.2 AutoCAD 2000i

Arguably, AutoCAD is one of the most venerable CAD platforms still in use today by designers and engineers worldwide. However, it is not until year 2000 that AutoCAD finally decided to jump on the Internet bandwagon. As we all may guess, the 'i' in the AutoCAD 2000i (A2Ki, in short) package stands for 'Internet.' Unlike Graphisoft's ArchiCAD for TeamWork, Autodesk, the vendor of A2Ki, did not go for a software solution for its users to implement teamwork management procedures in terms of design roles, privileges or other team-forming attributes. Instead, A2Ki provides a set of Web-based features that can be used to facilitate collaborative design in a loose way participated by multi-disciplinary team members. Here, a project team can be extended to include non-design professionals such as project clients or manufacturers of building products. We shall look at some of the major features of the 2000i release, focusing on how the Web and Internet have been taken up to evolve a traditional CAD package into a teamwork-promoting platform. For more comprehensive and detailed information about the software product, interested readers should consult other references on AutoCAD 2000i (see, for instance, Uhrskov 2000 and Stellman and Krishnan 2000).

Live Web Browser and Portal. The eye-catching commercial phrase used by Autodesk to promote the A2Ki release was that of "Internet-driven design," meaning that design will be driven by the Internet through the A2Ki platform. In fact, it is fair to say that the design of the A2Ki itself has been driven by the factor of the Internet. When starting the application, what users see is not that of a familiar AutoCAD drawing window but a Web browser. Assuming that the user's computer is in a constant live connection to the Internet, it first shows a Web page called "AutoCAD Today" containing latest information related to the design professions and building construction industry (Figure 7.2). Worth of mentioning in particular is that the Web browser can be customised by users to provide a personalised view of "Point A," an Internet portal service operated by Autodesk. Through the portal service, a design firm can set up a company's intranet running project-oriented Web sites as a means of sharing project information among project participants who may be constant on the move. Given the relatively long history of AutoCAD, to many users and CAD analysts the Internet-based live browser and portal seems to offer a refreshing feature. Although the current Web content provided by the Point A portal are specific to the US and Canada, further context-specific Web content could be developed in due course if more AutoCAD user communities in other countries find the Internet-based facilities beneficial.

Web Publishing and Net Conferencing. The A2Ki platform adopts still the long-lasting DWG file format. The browser plug-in "WHIP!" has been made available to view DWG drawing files on the Web. The "Publish to Web" facility provides users with a handy tool to publish DWG design data on the Web in either the DWF format (viewable with the WHIP! plug-in) or the JPEG image format. For uploading, users will also be prompted to specify the destination for project files on a network, intranet, or the Internet. This Web publishing feature effectively establishes Web-enable viewing of DWG drawings in AutoCAD, which can be useful for a project's team members to view each other's drawings conveniently even if they are located in different continents. Adopting Microsoft's NetMeeting technology, the "MeetNow" feature enables designers to conduct online project meetings across an intranet or Internet without leaving the AutoCAD environment. For this, Autodesk provides an ILS server (meetnow.autodesk.com) for users to log on and join a shared session of AutoCAD.

Drag and Drop. The Web-enabled drag and drop feature (or, "I-drop") is for a user to create design content on his or her own AutoCAD desktop by simply dragging content published on a Web page (in AutoCAD recognised file formats) and dropping them onto the user's existing drawing file. From a teamwork point of view, designers can post drawing content on their intranet sites and let other team members access and share the content by dragging and dropping from the project's Web pages. It is also envisaged as a productivity tool for speeding up the specification process: specifiers go to manufacturers' Web sites, which provide I-drop aware contents of their building products, and apply the drag-and-drop operations

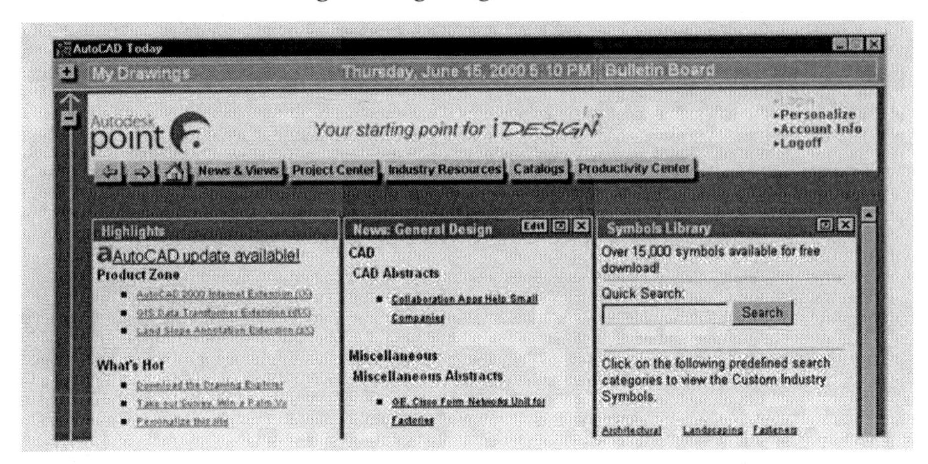

Figure 7.2 *AutoCAD2000i is opened with a customizable "AutoCAD Today" HTML-based window with a live link to Autodesk's "Point A" Internet portal.*

on the products they like to use. Based on AutoCAD's new Web-enabled desktop design tools, an I-drop object can be specified to provide a common gateway between a content provider's server and an AutoCAD user's desktop. However, while benefiting from the ease of browsing and generating design contents by I-dropping, it remains problematic for designers to keep track of changing states of the Web content in use. Consider a construction project producing hundreds of DWG files (a small size in today's standard), each of which contains a dozen or so I-drop objects, it will be a daunting task for the project team to keep checking manually whether or not the states of all I-drop elements are in synch with the latest Web content from which they were originally downloaded. What is required is a facility for generating and sending notifications automatically wherever and whenever there is a Web site update in which an I-drop object is involved. The notifications should provide sufficient information regarding the changes such that designers are assisted in reviewing and evaluating the implications of the changes on all the drawings where the I-drop objects in question appear.

Seen from the above Internet-enabled features, the A2Ki package has not been particularly promoted as a team-working platform by offering any specific organisational or managerial frameworks for design collaboration. We do not see particular roles that team members have to play in order to operate the drag-and-drop command, for instance. The connections made to the Internet or an intranet are generic, which do not lead to distinctive patterns of teamwork but rather open-ended collaboration and communication processes. Given that it is a relatively new platform at the time of writing this book, it is perhaps too early to say that the A2Ki approach is definitely going to mark a new beginning of the long AutoCAD product line. Nonetheless, one thing we can be certain about is the tendency of blurring CAD with the Web—a perhaps necessary next step for Autodesk and other commercial CAD software developers to take if the ultimate goal is to further integrate CAD with business-to-business and e-commerce computing.

7.3 MicroStation/J and ProjectBank

The DGN file format of the MicroStation platform we see toady was first developed by the Bentley Systems from Intergraph's IGDS (Interactive Graphics Design System) software design file format in the early 1980s. Then, in 1998, Keith Bentley, one of the co-founders and CEO of the Bentley Systems, Inc., published a white paper titled "ProjectBank" (Bentley 1998), which described a roadmap for developing a new generation of MicroStation platform. One of the fundamental ideas underlying the ProjectBank platform is to replace the traditional view of CAD as individual desktop/document processing with a more radical view of CAD as enterprise computing. Robert Aish, Director of Research at the Bentley Systems, once advocated the concept of "enterprise computing for construction," by which he meant that "design data is held in a sufficiently complete representation, and that changes to this representation are transactions that move the representation from one consistent state to another consistent state." (Aish 1999). *Transaction* is the key word in this statement, which has seldom been mentioned before in mainstream CAD concepts and principles. Quoting the "one million document" breakthrough in the new Hong Kong International Airport project, Aish questioned the current approaches to Construction IT and its applications in real world building projects—"Wouldn't it have been better to capture the design and construction process in fewer, more complete, representations?" (Aish 2000a).

Most of us would consider the increased productivity in producing drawings as the biggest gain from using modern sophisticated CAD systems. Designers working either as a solo or as a team member are capable of generating more drawings in less time. It is probably true that we now produce design and specification drawings for a construction project more than ever before, and this trend has certainly been fueled by the uses of CAD tools in design practice and construction industry. There may be very good reasons for design teams to produce more drawings in response to contemporary society's demands of more sophisticated built environments. However, there remains the question if we can equate the productivity of drawing production to that of project completion. Intuitively, no causal relations can be definitely drawn between quantities of drawings and productivity in construction projects (i.e., more drawings lead to more productive project completion, or, the opposite way around). But in statistical terms, we can be sure that it is more likely to have inconsistencies among a larger number of drawings than a smaller one. Seen in the one million more documents produced for the new Hong Kong Airport project, each of the documents is interrelated to many others in one way or another as same objects or spaces need be drawn and specified from different perspectives and for different purposes. Inconsistencies arise whenever the supposed interrelationships break down among documents due to human errors or mistakes.

Both Bentley and Aish consider the root of the inconsistency problem lies not in the resultant quantity of CAD drawings but in the way drawings are produced with existing CAD tools. Aish pointed out in particular the "desktop/document" metaphor adopted by the popular personal computing (PC) platform on which most CAD packages are deployed. Following the desktop paradigm, a 2D drafting application is essentially a replacement of a physical drawing board, which is used by designers to generate and edit discrete design documents. Each design document by itself is an incomplete representation and designers aim to achieve a more complete representation by producing a collection of documents as they see necessary. Working in this mode, as we can imagine, it seems inevitable that designers need to produce more and more documents in order to achieve a desired level of representation completeness as the size and complexity of a building project grows. An alternative way to better handle this situation, according to the Bentley's new CAD strategy, is to adopt the distinction between "definitive model data" and "derived report data" as pointed out by Aish (Aish 1999). Drawings in the definitive vs. derived universe are not discrete self-contained documents but *extractions* from the defining model data.

A simple example of this approach is constructing a 3-dimensional building model first and then extracting floor plans, elevations, and sections from the 3D model as drawings of the building. Notably, drawings are one among many possible forms of *reports* that can be derived from a definitive model. A major benefit of adopting this CAD strategy is that changes made onto definitive model data can be reflected immediately on derived report data; one can simply disregard the existing reports and extract a newer version from the revised model without actually modifying the reports. This is perhaps one of the most decisive conceptual devices that distinguishes individual CAD computing from enterprise CAD computing. The challenge for new CAD technologies is how to facilitate designers' achievement of better representation completeness while at the same time gaining a highest level of consistency among discrete design documents with minimal efforts. The development of the MicroStation/J and ProjectBank platforms seems to be a significant step taken by the Bentley Systems to realise this CAD strategy. Having discussed the conceptual background, we are now in a position to have a closer look into some of the system features provided by ProjectBank. Again, a comprehensive description of the platform is not intended here; for interested readers, more detailed and complete information on the software products can be found in other references (see, for instance, [Conforti and Sahai 2001]).

ProjectBank's Client-Server Architecture. ProjectBank DGN is a full-fledged client-server application run on Microsoft's Windows NT/98/2000 operating systems in conjunction with MicroStation/J. Clients and servers are connected through the TCP/IP protocol on a local area network. The server-side ProjectBank functions like a database transaction manager that receives and processes requests for data from clients. A server provides three major storage areas: schemas, components/objects, and component history (see later for more discussion of each area). The client-side ProjectBank is manifested in "project briefcases," which are created on local seats of MircoStation/J against a particular project defined in the project registry on a server.

Once created, a briefcase can hold multiple working design files that are local copies of (user-selected) files on the server's data store. There can be several projects set up on a ProjectBank server, and a team member can open up any number of briefcases for the project he or she is working on. ProjectBank's client-server architecture can be configured to meet different project needs: a single independent designer (both server and client are on the same workstation); a centralised workgroup (several clients connected to a single server); a distributed project team (one or several clients connected to a hierarchical network of servers). More importantly, the client-server architecture provides the foundation to build up a novel workflow technology in ProjectBank (see below).

Elements, Components and Schemas. Related to the aforementioned strategy of a single defining building model and derived data, Bentley puts forward a process migration from geometry-based drafting or modelling to Engineering Component Modelling (ECM). On MicroStation, designers represent design intent in terms of plain geometric entities such as lines, B-Splines, arc, and rectangular block. ECM is a new technology for incorporating "comprehensive computable model of design intent," to use Aish's terms, on top of geometric data so that designers can communicate with machines in terms of "components"—data types with some content semantics (doors, windows, floors, for instance). ECM is made possible through defining "schemas." A schema is a group of classes (as in object-oriented programming) that together define component types used to model a specific construction or engineering function (Bentley 1998). The JMDL (Java MicroStation Development Language) was developed by Bentley for designers to create component-defining schemas. This is where MicroStation/J comes about (the J is for Java). As an off-shelf package, the current release of ProjectBank comes with a ready-made DGN schema that can map design elements in a traditional MicroStation design file to instances of DGN Schema components in ProjectBank where the persistent attributes and behaviours are stored. Using JMDL, there can be many schemas defined on a ProjectBank server. In fact, Bentley has planned to provide a DWG schema (ProjectBank DWG) in their release of ProjectWise. It is expected that third-party developers committed to ProjectBank technology will be developing their own domain-specific applications on the basis of JMDL.

Commit and Synchronise. All working design files residing in a project briefcase are accessible only to the owner of the briefcase. A designer's creating and editing design files are kept local to each of the briefcases. For other workgroup members to share the edited materials, he or she needs to "commit" the briefcases back to the server. For group members to receive colleagues' new designs or design changes, they need to "synchronise" their briefcases with the server. Both commit and synchronise are specific menu-driven commands in ProjectBank that users can initiate at any time. If there is something to be synchronised in a given state of a user's briefcase (as detected by the ProjectBank's transaction manager), the commit operation on the user's workstation will be disabled to ensure that the policy of *synchronise first and then commit* is followed. It should be noted that every committing operation involves executing the associated schema; that is, every element in a briefcase design file is examined and a corresponding schema component is created on the ProjectBank server, and elements can be recreated at any time should the designers decide to extract or derive the design file back from ProjectBank (Sahai 2000). Of course, the exact nature of an extraction or deriving depends on what schema is in action. One can think of defining a schema for the purpose of performing building energy performance simulation: An architect creates a design proposal in her briefcase and commits the design to ProjectBank through the building energy schema. A building energy specialist comes along and extracts the design proposal from the Bank to analyse the design proposal by calling upon his favoured simulation program. Through ProjectBank, it is therefore possible for the architect to work on a design file format that is different from the one required by the energy engineer, yet they can collaborate with each other on the common ground established through the energy schema. However, it is an open question whether or not we can always create a sound schema between any two applications that can deliver complete data mapping.

Transactions, Component History and Concurrent Editing. A transaction in ProjectBank is a record of the content difference between two consecutive actions of committing a briefcase to the server. The differences may be some design elements being deleted, added, or translated, transformed, or a combination of these. ProjectBank keeps a history of all the transactions (at the component level) as committed by any group members since the beginning of a project. There are no hard-coded rules regarding how frequent users should commit

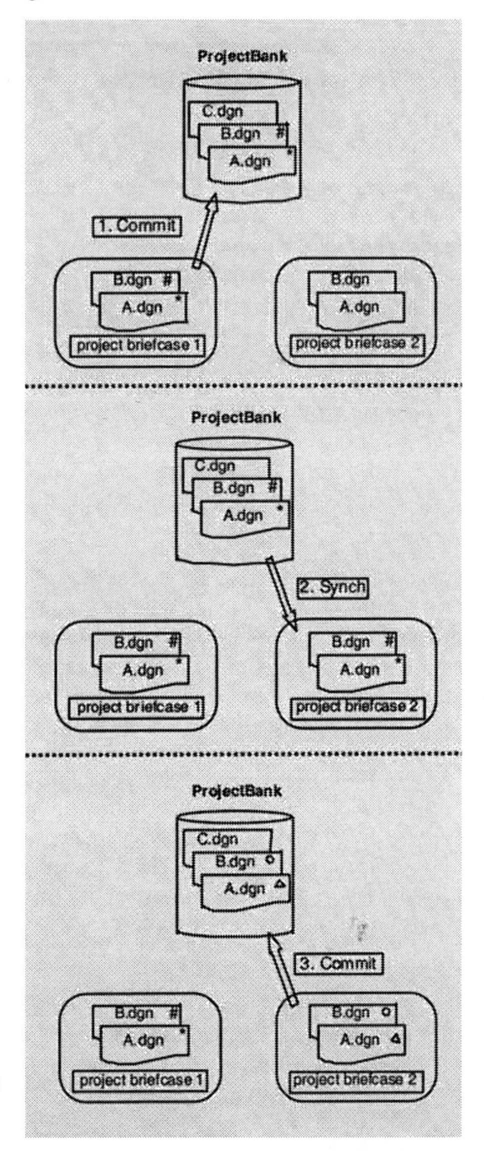

Figure 7.3 An example of workflow in ProjectBank involving "Commit" and "Synchronize" between two designers. The designer using project briefcase 2 must synchronise first before commit change.

transactions; it is entirely up to the individuals and the design teams who may set up routines of making transactions. Because of the historical records maintained, all transactions in ProjectBank are reversible, that is to say, a design file can be reversed to any previous state recorded and all the alterations made in between are disregarded. In ProjectBank's terminology, this is called "Revisions." An automatic versioning mechanism is provided for every component known to the Bank. The Revision Explorer is the graphical user interface for team members to browse a project's revision history, finding out who made what changes at what time regarding what considerations. Because of the functionality brought by transactions, component history, and revisions, ProjectBank allows editing of the same design file(s) by different group members *concurrently*. As seen before, this is not possible in the case of ArchiCAD for TeamWork. Aish once characterised this facility as "Optimistic Long Transactions," which are considered viable and desirable for three reasons (Aish 2000a):

- Not to create a complete blockage to parallelisation of design;
- Designers usually coordinate not to make conflicting changes to the same data items;
- A technology to resolve conflicts is made available should they occur.

Resolving Conflicts. Members of a workgroup will not find any conflicts when working with their project briefcases. It is until the moment of a participant commits a briefcase that conflicts may occur. "Change merging" is the computational process in ProjectBank to detect if there exists a discrepancy in the same component instance that appears in an already committed design file and in another (same) design file that is about to be committed. If no discrepancies are found, the changes are deemed compatible, and change merging takes place automatically, otherwise the changes are conflicting and the later committing will not be executed. A notification of conflict will be sent to the workstation where the committing is initiated. Conflicts can be resolved by either (a) further briefcase-based editing so that changes become compatible with the committed version, or (b) going through a revision session and reverting the committed file to a previous state against which the current proposed changes become compatible. ProjectBank provides conflict detection and notification capabilities, but the actual resolving of any conflicts occurred due to concurrent editing remains in the hands of human designers who may need to work out how to coordinate or negotiate their differences through virtual or face-to-face meetings.

From the system features reviewed above, we may conclude that parallelisation of design is considered in ProjectBank a positive opportunity for collaborative design that can be dealt with in an optimistic strategy. By contrast, a pessimistic approach would be to impose exclusive locking of design elements, areas or files (as seen in ArchiCAD for TeamWork) to avoid potential conflicts altogether from the outset. Bentley's current releases of the MicroStation/J and ProjectBank platforms have not really exhausted all the technological possibilities that their researchers and developers have envisaged. It will be interesting to see empirical studies to be carried out to evaluate the impacts of the CAD as enterprise computing approach on real world design practices. As already being pointed out by Aish, to really reap the benefits of employing a pro-enterprise strategy and taking up new technologies like MicroStation/J and ProjectBank (-Wise), "process re-engineering" for collaborative design is a crucial issue to be addressed by design practitioners and the AEC industrial sectors (Aish 2000a). It could be argued that in fact the approach of defining and deriving in building design modelling has not been a total stranger to architectural designers and engineers in the past. The creation and uses of the funicular modelling in the Colonia Güell Church design project is a case in point (Section 5.1, Chapter 5). Today's "re-engineering" issue seems more to do with reflecting on what CAD technologies have been developed and marketed, and on the other hand, how these technologies have been perceived and used in practice.

7.4 XML-based schemas for pan-industrial information exchanges

It is true that with the teamwork supporting features currently provided by the commercial CAD software collaborative building design can take place in various forms. Most of the collaborative features, as reviewed above, have been developed as add-ons to extend traditional CAD capabilities so that they can be operated as multi-user drawing or modelling platforms. ArchiCAD for TeamWork, AutoCAD2000i, and MicroStation ProjectBank DGN all put up collaborative design facilities according to their rather different strategies of how teamwork should be managed within the applications. To a large extent, the degree of success of these products will hinge on the assumption that it is suffice to use only these CAD systems in collaborative building design. If we define the scope of collaboration as collaborative editing of production drawings, these industrial packages may be sufficient on their own. But if a more extensive scope of collaborative design is considered, there are further important issues yet to be addressed in CAD software development. As we know, a building design project often involves more then one type of professional group in forming a project team (engineers, architects, quantity surveyors, project managers, to name a few), and many different software applications are used by the sub-groups in fulfilling their different project roles. Supporting design collaboration to its fullest extent will need to address the difficult problem of how project data can be created and accessed across the different software applications used by member groups of a project team. Powerful CAD tools are best for designers to produce 2D or 3D representations of built forms, but these are one among many other types of software required for a building project.

Since 1999, several initiatives have been launched by the building design and construction industries world-wide to develop XML-based schemas for pan-industrial data communications and information interchanges over the Internet. Before the XML-based approach, there have been other attempts to tackle similar issues concerning data standardisations or interchanges, but now many believe that XML-based schemas will become the most influential frameworks ever proposed on the ground that wide acceptances of these schemas by the industrial, academic and government users alike are forthcoming. Building on the strength of the Web, developments of XML-based data description schemes may affect construction IT in two major aspects: (a) existing software applications may be extended to accommodate XML-based data schemas such that heterogeneous applications can talk to one another over the Internet; (b) new data modelling and communications technologies may be developed according to some newly specified XML schemas. In either way, new possibilities of multi-disciplinary, project-wide collaborative working in building design and construction may be opened up on a scale unseen in the current HTML-based technologies. Up to date (mid-2001), there are five working groups organised to develop pioneering XML-based data structuring frameworks specific to building design and construction: aecXML, bcXML, gbXML, eBuild-XML and MasterBuilder Construction Management and Accounting. Before introducing some of these specific schemes, we need to look at the XML (eXtensible Markup Language) itself as the foundational technology.

The Extensible Markup Language was first specified and reached a Recommendation status by the World Wide Web Consortium (W3C) in February 1998 (W3C XML Working Group 1998; Bray et al. 2000). The XML is intended as "the universal format for structured documents and data on the Web." The design of the XML has its root in the Standard Generalised Markup Language (SGML) founded back in 1986 (ISO 8879) from which the enormously popular HyperText Markup Language (HTML) was also spawned. Both HTML and XML belong to the family of electronic data tagging (marking-up) languages. But the similarity between the two ends here. We use HTML to publish hypertext content on the Internet as Web pages. Basically, the set of tags specified in HTML are used to format how Web content should be presented. XML, by contrast, was conceived to enable Web pages of separated content from presentation (appearance). In short, HTML pages are designed for the human eyes, while the XML format is for advanced machine processing. For the over one billion HTML Web pages now publicly accessible on the Internet, the only processing applications developed for the Web pages are search engines that locate and rank registered HTML documents. With XML-enabled Web pages, far more information processing applications can be developed. This is

HTML	XML
<html>	
<head>	
<title>My Book</title>	<?xml version='1.0' ?>
</head>	<book title>Weaving the Web</book title>
<body>	<author>Berners-Lee, Tim</author>
<p>Book Title: Weaving the Web</p>	<publisher>Orion Business</publisher>
<p>Author: Berners-Lee, Tim</p>	<year>1999</year>
<p>Publisher: Orion Business</p>	<city>London</city>
<p>Year: 1999</p>	
<p>City: London</p>	
</body>	
</html>	

Table 7.1. A simple example of HTML vs XML.

possible because a new approach to creating data tagging has been established by XML. Unlike HTML tags, XML tags are used to lay type and structure over information, which are defined to describe the "meanings" of the data. More importantly, XML puts control of the tags set in the user's hands; user groups can create new tags as needed (Usdin and Graham 1998). This is clearly not the case in the HTML tag set, which is under the control of the developers of HTML browsers and no end-users are in a position to add new entries to the set of tags.

Let's first consider a somewhat trial example of comparing a piece of information written in HTML and XML respectively (see Table 7.1):

The HTML example above (left) is a well-formed HTML page (file) viewable in any HTML-compatible browser and it can be located on the Internet once it is registered with some popular Web search engine sites such as Yahoo or Google. The XML example (right), on the other hand, can also be displayed in a Web browser (shown in a single line instead with all the tags, i.e., word or phrases contained in pointed brackets, ignored). The XML tags offer no clues as how the information should look on a browser or on paper, but more useful things can be potentially done with the formatted information: (a) It allows for different renditions of the same information (on a PC, PDA, or a mobile phone); (b) It allows for more intelligent queries such as "Find `the <book title>` with `<author>` Berners-Lee and `<publisher>` Orion Business in `<year>` 1999; (c) Two or more software applications may exchange this simple XML record, each program can "understand" the exact information stored within a specific element, such as the data in `<book title>` "Weaving the Web" (i.e., the title of a book); (d) an XML hyperlink, not shown in this example, can open up a menu of several options, such as a graphic image of the book cover, a Java applet to order the book online, a message box

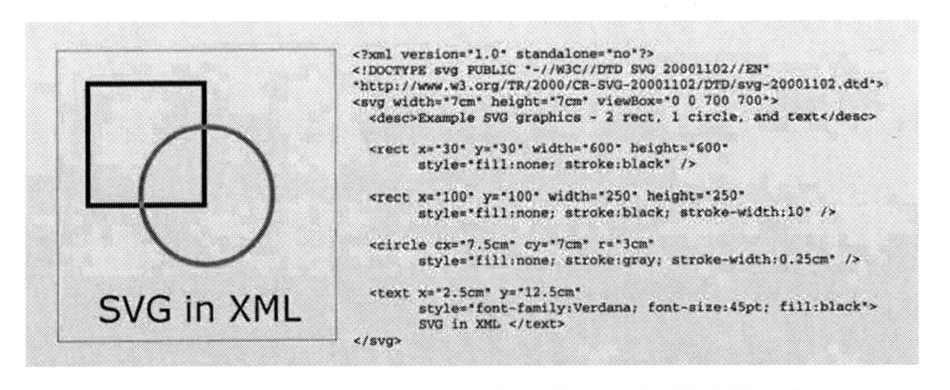

Figure 7.4 (above) A simple drawing specified in a piece of Scalable Vector Graphics (SVG) code.

Figure 7.5 (right) Applying the "Find in SVG" operation onto a SVG map to locate the "Theater Row" text element on the map. All text elements in a SVG drawing can be searched against user input query; the SVG Viewer highlights all elements matching the query terms.

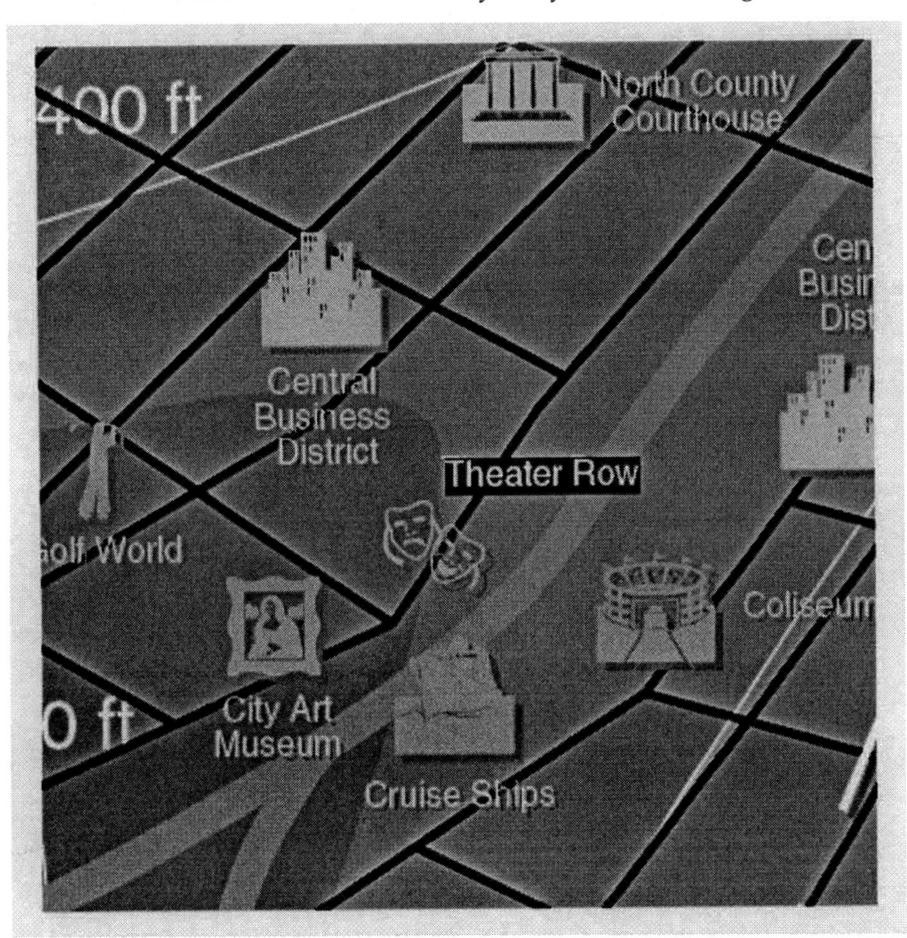

Find in SVG

Find what: Theater Row

Find Next

Cancel

Match whole word only

Direction

Up   Down

Match case

containing latest reviews of the book, or a page with more links to the author's other books. However, it should be said that accomplishing (a) to (d) depends on the various software applications made available. XML does not tell how the data should be processed, which is considered as the obligation of applications software to accommodate the tags specified. To give a more substantial example of XML, Figure 7.4 presents a simple drawing specified in the Scalable Vector Graphics language.

The Scalable Vector Graphics (SVG) is an open-standard data format for describing two-dimensional graphics in XML. The tags shown in Figure 7.4 are specified in the SVG's Document Type Definition (DTD) published by the W3C SVG Working Group. The actual graphic image is displayed by Adobe® SVG Viewer installed as a plug-in on Microsoft Internet Explorer. The same SVG code can also be read by Batik SVG Viewer, which is developed by the Apache Batik project employing the Java programming language.[36] A closer look into the SVG code, one can think of an application that will allow a viewer to query the image, asking questions like "What is the area of the rectangle?" "What is the distance between the centres of the rectangle and the circle?" "What is the area of the intersection of the rectangle and the circle?" etc. Borrowing the "Printable Map" demo from Adobe's SVG Web site, Figure 7.5 shows an example of applying the "Find in SVG" operation onto the SVG map image, which highlights the "text" element found in the map by matching the words entered by the viewer.

These querying operations are not applicable to HTML-displayed images that are usually presented in some compressed bitmapped formats (e.g., JPEG, PNG, etc). Apart from being an open-standard, another notable benefit of the SVG format is its economy of data transmission over the Internet. Using the export facility in Batik SVG Viewer, the same code (1K in size) results in a JPEG file of 9K.[37] Using SVG defined images as an example, we may foresee the potential interoperability of XML described and structured information. The increased ease of information sharing and exchanges among domain-specific software may thus open up new routes for supporting human users' cooperative working. However, this is not the place to delve into further details regarding XML and SVG. For more comprehensive descriptions of XML and SVG developments, interested readers should consult other references (see, for example, [Harold and Means 2000; Ferraiolo 2000]). Having seen these XML examples above, we shall continue to look at building design, construction and operation related XML schemas that have been proposed and developed by various working groups: aecXML, bcXML and gbXML.

aecXML

The aecXML is a framework for electronic communications in the architecture, engineering, and construction industries. It provides an XML-based schema to describe and structure the information specific to the participants involved in designing, constructing and operating buildings, plants, infrastructure and facilities (aecXML 2000). The aecXML Working Group, first instigated by Bentley Systems, Inc. in August 1999 now administered by International Alliance for Interoperability (IAI), consider aecXML a "transport mechanism" by which information can flow seamlessly among various software tools used by AEC professionals over the lifecycle of a project. Not to produce yet another neutral CAD data format, the working group aim to identify a set of keywords and named data attributes that may be accepted by the group members as common naming logic and data grouping. Subsequently, a preliminary specification of aecXML was submitted by Bentley Systems, Inc. as the initial proposal of XML grammar for AEC data communications on the Internet.[38] The types of information addressed by the current schema include: Documents (drawings, specifications, contracts, purchase orders, etc.), Projects (design, construction, operation & maintenance etc.), Building Components (items from a building product catalog, assemblies etc.), Professional Services and Resources (architects, engineers, contractors, suppliers, etc.), Organisations (standard bodies, government agencies), and Software (CAD, cost estimating, project management, etc.). As the initial proposer of the schema, Bentley Systems has explicitly expressed that aecXML is a non-proprietary framework subject to continuous development by public participation (Newton 1999).

On the basis of the aceXML initiative, Bentley Systems has moved to develop two collaborative commercial projects: one was to ally the MicroStation platform with the McGraw-Hill Companies' Sweet's Group (www.construction.com), a provider of a large online database of building production information; and secondly, to integrate MicroStation TriForma with Means CostWorks, an online construction cost estimating service provided by the R. S. Means Company, Inc., (www.rsmeans.com). With the former services alliance, they aimed to provide users of MicroStation with simultaneous access to relevant product information such as 3D specifications, CAD details and local sales contacts during project design and development. In the latter integration, it is expected that designers will be assisted with automated access to regularly updated cost information not available through other means and project clients will consequently receive more accurate quantity and cost information throughout the design process.

bcXML

The Building Construction Extensible Markup Language (bcXML) was a pan-European initiative funded by the eConstruct project under the European 5th Framework research programme. The project aims to develop and disseminate bcXML as a versatile but low cost communication infrastructure accessible to the building construction industry in Europe.[39] Started in January 2000, with partners from UK, Greece, Holland, France, Germany, and Norway, bcXML is expected to play several roles in improving the performance of European building construction industry: (1) facilitating e-business processes among clients, architects, engineers, suppliers and subcontractors; (2) integrated with e-commerce and design/engineering applications; (3) supporting virtual construction enterprises over the boundaries of the individual European member states. As identified by the bcXML Working Group, there are two major issues to be addressed by creating the pan-European framework: one is the different (natural) languages and the other is the different classifications of project information used by European state members.

```
<?xml version="1.0" ?>
<gbXML xmlns="http://idea-server.com/xml/gbxml/0-31"
xmlns:xsi="http://www.w3.org/2000/10/XMLSchema-instance"
xsi:schemaLocation="http://idea-server.com/xml/gbxml/0-31 http://idea-
server.com/xml/gbxml/0-31/GreenBuildingXML.xsd" id="x1" engine="DOE-2.2"
lengthUnit="Feet" areaUnit="SquareFeet" volumeUnit="CubicFeet"
temperatureUnit="F" useSIUnitsForResults="false">
<Campus id="c1" designHeatWeathIdRef="we1" designCoolWeathIdRef="we2">
  <Name>Default Minimum Project</Name>
  <Location>
  <Name>Sonoma</Name>
  <Elevation>500</Elevation>
  <ZipcodeOrPostalCode>94589</ZipcodeOrPostalCode>
  </Location>
  <Building id="b1" buildingType="Office">
  <Name>North Bldg</Name>
  <Area>5768</Area>
  <Space id="sp1" zoneIdRef="z1" lightScheduleIdRef="s3"
equipmentScheduleIdRef="s3" peopleScheduleIdRef="s4">
  <Name>NE Zone 1</Name>
  <PeopleNumber unit="NumberOfPeople">3.85333</PeopleNumber>
  <TotalPeopleHeatGain unit="WattPerPerson">400</TotalPeopleHeatGain>
  <LightPowerPerArea unit="WattPerSquareFoot">3.8</LightPowerPerArea>
  <EquipPowerPerArea unit="WattPerSquareFoot">1</EquipPowerPerArea>
  <Area unit="SquareFeet">1156</Area>
  <Temperature unit="F">74</Temperature>
  <Volume>10404</Volume>
  </Space>
.....
```

Table 7.2. a portion of gbXML code for a small office design by GeoPraxis.

Still much under development as this book went into press, bcXML is being defined to support national language and classification specific views on construction project information for project partners of different member states. But that specificity will be delivered through bcXML's classification-neutral object identifications. According to Tolman and others (Tolman et al. 2000), object identification has to do with a "common understanding of what constitutes an object" (i.e., what makes a column a column?); and bcXML will be built upon LexiCon, a large and ordered set of building and construction concepts with attributes established by another European project CONCUR. Ultimately, bcXML will support multiple European languages and definition systems like taxonomies, classifications and dictionaries enabling national front-ends for bcXML-data access (Böhms et al. 2001).

gbXML

The Green Building XML Schema is specialist initiative on a smaller scale, which is designed for architects, building designers, CAD developers, and product manufacturers who want to incorporate green building principles in their designs, tools, and products. More specifically, it will facilitate the transfer of building product characteristics and equipment performance data between manufacturer's databases, CAD applications, and energy simulation engines. gbXML seems to be founded through the sole act of GeoPraxis, Inc., a provider of XML-based building energy and environmental analysis software. GeoPraxis published the gbXML XML Schema on the Internet in August 2000 [GeoPraxis 2000]. The following shows a small portion of a gbXML file created by GeoPraxis as an example for a small office design[40] (Table 7.2).

Gaining support from the California Energy Commission's Public Interest Energy Research (PIER) Program and CAD developer Artifice, Inc., GeoPraxis' first major implementation of gbXML Schema is the Energy Analysis Module (EAM). The idea is that EAM will receive data from Artifice's DesignWorkshop 3D CAD tool (or, any other 3D CAD software supporting gbXML) and generate a well-formatted input file for conducting DOE-2 building energy simulations. With the DOE-2 simulation engine connected, designers will be enabled by EAM to evaluate energy cost of a commercial or residential building design at an early design stage. The "what-if analysis" capabilities in EAM will allow designers to investigate energy-related impacts of their designs against a range of factors such as climate, orientation, building operation, construction materials, and fenestration layout (we may see some of these from the gbXML code shown above). Given that the original DOE-2 program has not been an easy tool for architects and designers to operate without special training, it is expected that tools built around the gbXML Schema may bring the technical expertise of building energy analysis within a range that designers may find easier to access.

As in April 2001, the well-known Robin Cover's list of proposed standards and industry initiatives in developing community XML tag sets on the Web (http://www.oasis-open.org/cover/xml.hml) shows other two building-related XML schemas: eBuild-XML and MasterBuilder Construction Management and Accounting. Because they are not directly concerned with design processes (one for purchase and invoice exchanges among major house builders in the UK, and the other for construction cost/project management in the US), we shall not discuss them here. It is likely that more XML-based schemas will be identified and evolved in the future that look into other domains of information structuring special to building design, construction and operation. In general, the number of dedicated working groups in creating shared XML schemas has increased significantly. Cover's listing was quoted by Usdin and Graham in September 1998 and it showed 74 entries (Usdin and Graham 1998), while in May 2001, the list of entries increased to 439. As Usdin and Graham commented, it is through the efforts of these working groups that the most expensive, complex, and probably most important work in making XML viable is being done. However, we should not think that all these XML schemas are one-off efforts. Like software developments, XML schemas may go through evolutionary versions as seen needed by the host working groups and other participants.

From what we see in the above, XML is by no means a rocket science and technology. To many observers, the strength of XML will be decided by its global rapid acceptance rather than by its technological advancement. The success of HTML could be a precedent to XML in this respect. However, XML by itself does not accomplish the goal of data sharing and exchange on the Web. Much of it depends on wide public participations in developing shared XML schemas and processing applications that will exploit XML-tagged information. But we should be aware that XML is inherently far more ambitious than HTML as it attempts to deal with the difficult realm of laying semantic structures (or, meanings) over information and human knowledge. We have briefly touched on this issue at the beginning chapter regarding non-prescriptive design computing. From past experiences of knowledge-based design systems, prescriptive technologies could result from dealing with semantics or meanings of information in a closed universe in which some individuals or organisations stipulate what is meant by a 'column', for instance, and everybody else ought to stay with it in order to talk about or draw a column. Will XML schemas end up with a similar effect of 'prescriptiveness' that is particularly problematic in creative design communication and collaboration?

It can be argued that we now face a very different situation due to the Internet and the Web, which may provide an open universe for introducing shared semantic structures over information presented as Web pages. Tim Berners-Lee and others call this the Semantic Web, which can be developed by three basic components: XML, Resource Description Framework (RDF) and Ontologies (Berners-Lee et al. 2001). Using XML for its syntax and Universal Resource Identifier (URI) to specify entities, concepts, properties and relations, a RDF provides a built-in mechanism for expressing meanings of information in a form that computers can readily process. Ontologies are like reference books stored in public libraries, defining relations between concepts and logical rules for reasoning about concepts. Software applications are said to "understand" the meaning of semantic data on a Web page by following links to specified ontologies. It is envisaged that the Semantic Web, if properly developed, can facilitate the "evolution of human knowledge as a whole." Perhaps, there is a new prospect following the Semantic Web that we will be able to perform knowledge-based information processing and communications without being subject to prescriptive technologies eventually.

In the field of building design, construction and operation, we begin to see industry-led initiatives to embark on XML-based frameworks for cooperative processing of project information, which is not possible to achieve through HTML-based Web content. But creating XML schemas is not the end of the story according to the Semantic Web theory. We have not seen ontologies written with regard to XML schemas or frameworks. If the XML-based approach continues to mature and to be accepted by wider user communities in developing the Semantic Web, we can envisage the beginning of a new era in which the software tools used by participants of a building project may not be large expensive systems sold by established commercial developers. Instead, the software programs will become much smaller in size but more capable of processing information with special methods or from particular views and of collaborating with other small programs in real time over the Web. Open data schemas may gradually replace proprietary CAD data formats and thus open up opportunities for more distributed development of innovative software applications. In the long run, this new technological trend can stimulate new ways of thinking how computer-supported design collaboration should be implemented in practice.

Summary

We discuss the teamwork supporting features currently offered by some of the most popular industrial CAD platforms: ArchiCAD for TeamWork, AutoCAD2000i and MicroStation ProjectBank. Using TCP/IP on Microsoft Windows or Apple Macintosh OS, ArchiCAD for TeamWork, developed by Graphisoft, provides a simple file sharing facility for online or offline design collaboration. A Team Leader first sets up and creates a central master project file and then makes it accessible to other designers by signing in as Team Members. An individual Workspace is created by a Team Member reserving an area of work on the central project file through a number of data designating tools such as Layers, Stories, and Polygon Marquee. All elements reserved in a workspace are locked and can only be edited by the owner of the workspace. The TeamWork software is able to detect overlapping of working areas/elements between any two individual workspaces, which is not permitted. This exclusive locking policy is to ensure that inconsistencies among the drawings data created by team members may not occur. A team member can release his or her reserved area at any time by signing out so that another member may continue to work on it. The TeamWork platform assumes that a clear division of project work can be pre-assigned by a team leader or it can be done by team members themselves collaboratively.

AutoCAD2000i, developed by Autodesk, provides the AutoCAD platform with a Web-enabled desktop to achieve improved communication throughout participating organisations of a project, including clients, suppliers, vendors, and other project team members. The application is opened with an "AutoCAD Today" Web page, pointing to Autodesk's Point A Internet portal, which can be personalised to display project-specific information. The "Publish to Web" feature provides an easy way of converting a DWG file into a well-formed Web page containing the drawing in DWF or other Web-ready image formats. The WHIP! plug-in is provided for viewing DWF drawings on the Web. The "I-drop" technology is perhaps the most innovative feature, allowing users to drag-and-drop AutoCAD recognisable content published on the Web such as manufacturers' product specifications or team members' drawings. Other teamwork supporting features include MeetNow (for hosting virtual meetings on Intranet & Internet) and Live Updates (for delivering software updates over the Internet) that make use of existing Microsoft technologies.

Developed by Bentley Systems, MicroStation ProjectBank is a full-fledged client-server application deployable with the Windows NT/98 operating platforms as a company's intranet. Together with MicroStation/J, ProjectBank presents a significant change in the company's CAD strategy, moving from individual computing to enterprise computing. Fundamental to the pro-enterprise strategy is the Engineering Component Modelling framework for designers/engineers to create schemas, components or objects comprising a project. ProjectBank's server-side application provides storage areas for storing schemas, components or objects and component histories on a per project basis. For a large multi-firm project, it is possible to deploy a hierarchy of ProjectBank servers that for a project-specific network. The client-side application is manifested through team members creating individual Project Briefcases on local workstations. Designers work on copies of design files derived from the project bank, and concurrent editing of the same design file by several designers is allowed. Participants make design contributions by "Commit" briefcases back to the servers. "Change Merging" at the component level will only take place if the client copy is deemed compatible with the existing original file on the server. If changes are conflicting, designers can call on "Revisions" to review and resolve the conflicts. Concurrent editing and conflicts resolving are viable because of the "Component Histories" stored in ProjectBank that enables reversible editing. Believing in *optimistic long transactions* that designers usually do, ProjectBank considers *parallelisation of design* is an opportunity to enhance collaborative design rather than something to be avoided from the outset.

Finally, we look at several industry initiatives of developing XML-based schemas for Web-enabled data exchanges between software applications specific to building design, construction, and operation. Now widely considered as a successor to HTML, XML is being used to create community XML tag sets and schemas that will give "meanings" to Web content that applications software can readily process. By subscribing to the same XML schemas, it is possible for computer programs developed in different domains of applications to talk to each other and deliver cooperative processing of semantic data expressed as Web pages. Currently, three working groups have published XML schemas targeted at building design applications: aecXML, bcXML and gbXML. Following these pilot schemas, initial collaborations between different commercial packages have taken place: MicroStation with Construction.com and Means CostWorks, and GeoPraxis' Energy Analysis Module connecting Artifice's DesignWorkshop with the DOE-2 building energy simulation engine. If the XML-based approach to data sharing and interchanges over the Internet continues to expand and mature, it is expected that industrial proprietary CAD data formats may be one day replaced with open-access XML schemas and ontologies, which may thus open up new possibilities of Web-enabled collaborative design.

8 Futures of Groupware for Collaborative Design

One of the major developments of CAD in architectural design will be the functionality and interfaces that are capable of supporting multidisciplinary designers and other project participants working as teams. In the immediate future, we are likely to see that computing communications as a teamwork-supporting facility is an integral part of a CAD environment. However, the richness and complexity of human interaction in collaborative architectural design still presents considerable challenges to the development of design computing systems that will be adequate to sustain group dynamics in collaborative modelling of architectural design. Since its inception, the development of computer-aided design in architecture has been through various stages. I have briefly described the historical developments of CAAD in Chapter 1. Clearly, each stage of how we defined and implemented CAAD strategies and techniques has been influenced by the inventions in computer science and information technology put forward at those times. Without a doubt, architectural CAD system development in the near future will be substantially influenced, if not directly engineered, by the developments in computing communications. Networking capabilities of connecting with project-wide databases, the Internet, intranets and extranets will be essential to any major computer-aided design platforms. More importantly, the society at large shall demand such development as more and more people are experiencing many fundamental changes in their daily lives due to digital and networking technologies.

However, on the other hand, the input of knowledge of human factors to the technological developments will continue to demand our better understanding of interpersonal and group communication and collaboration activities. Design and particularly architectural design is a rich and complex subject area for eliciting requirements in the making of new design tools and in evaluating existent system assumptions or boundaries. I have presented a number of case studies of building design projects, examining the properties of the design representations and the implications for design communication for teamwork. It goes without saying that there can never be enough studies regarding how designers work cooperatively as design teams, and it is important to disseminate and share the research findings with the system design and development communities. The final chapter of this book presents a survey of some topical areas for further research and development of computer-supported collaborative design in practice and in design education. Better understanding of cooperative design processes and activities remain crucial to the continuous development of usable software and hardware components. More experiments are needed to try out mechanisms and interfaces for information sharing, design development coordination, and synchronous/asynchronous remote collaboration.

8.1 Toward a hypermedia case bank for design studies

Several case studies of team working in architectural design on the basis of observing what designers generated in representing their design concepts and how the representations may have been used in communication and coordination for design development. The individual cases as studied, seem to suggest that there are patterns of team working in design that we can attribute to the activities of collaboration. Following the teamwork patterns analysed, I was to be able to better characterise and describe what was involved in collaborative architectural design; the descriptions and characterisations were then elaborated with regard to describe how computer-based collaborative design environments may be defined.

A problem of research methodology

However, it should be pointed out that the so called teamwork patterns are *abstractions* of real world events, and any abstractions run certain risks of either oversimplifying the complexity and variety of the real world or simply being *reductionistic*. The bottom-up and top-down scenarios of team working discussed in the preceding chapters are at best two abstractions of what might have happened in reality. I believe that if the scope of the case study is broadened to conduct more case analyses, it is likely that further patterns of teamwork may be identified and present issues and requirements for computer support not yet explored. In pursuing practice-pulled research, the purpose of identifying teamwork patterns and analysing in greater detail the components of the patterns is to produce primarily consistent programmes for system research and development. In this regard, the collecting of field data in conducting design process or activity studies is of paramount importance. There will always be the conundrums regarding what design activities to study and how many design studies are sufficient to inform a workable system development programme. Also, are studies of well-known architects and designers at work more accurate or representative of the studies of, say, design students in an educational setting? There are no easy answers to these questions. Simply, as far as the quantity of design studies is concerned, there are no clear lines to which we may mark as the ending of design studies and the beginning of system developments.

More research into the features of teamwork patterns will be valuable to the development of next generation computer support for collaborative design. From a practice-pulled viewpoint, computing system designs are expected to respond to the requirements elicited from design practice. Assuming ideal system development results, the systems engineered should behave supportively of certain patterns of collaborative design work. But does this imply that the systems will dictate how designers should collaborate with one another by adopting particular teamwork patterns? Or, does it matter that a system supporting the top-down approach, for instance, may appear completely useless to design teams working in the bottom-up way? Like any other computing tools, collaborative design support systems are tools devised to serve certain purposes. The question is how general, flexible, or specific otherwise, a system should be in order that a majority of the end-user community will find the tools useful and cost-effective. Without saying, research on teamwork patterns alone is not sufficient to inform comprehensive system production strategies. We may think, for example, a collaborative design environment large enough to accommodate more than one teamwork patterns;

the design team can decide which pattern suits best the design project at hand and then switch on a particular mode of system support.

Drawing on research into teamwork patterns, the most general property we can describe about system requirements is perhaps the *flow of information*. From an architectural design viewpoint, I found that describing information flows in collaborative design is a good way of clarifying the dynamic relations between design parts and wholes as generated and manipulated by members of a design group. The flows of information in the top-down and bottom-up patterns suggest different approaches as to how parts and wholes are formed and how changes made by participants in parts or wholes affect the evolution of collaborative design.

In retrospect, since its inception in the 1960s, research into what designers do and how designers think have generated rich resources for better describing and understanding design activities and processes. Design processes research may have not fundamentally changed the ways that designers think about and make buildings; however, we do have theories and concepts as put forward by researchers that serve as intellectual vehicles for exploring more deeply the spectrum in between idiosyncrasies and rational (scientific) models of design. And it is certainly more evident that design processes research has played a part in the evolution of components and infrastructures in computer-aided design system development. As the information technology continues to evolve, collaboration between system developers and design process researchers can be ever more important in producing responsive and innovative tools in support of aspects of design practice.

The construction of a hypermedia-based case bank

Given my previous experience of conducting studies on the case histories, I consider that there are areas for improvement methodologically. In particular, observations of how designers think and work can be documented in digital media that will allow for structured organisation of the raw data elicited from design studies. Conducting and presenting design process research in a digital form can deliver benefits in at least three aspects: seamless capturing of design activities, readiness of data collection for processing and retrieval, and versatility for interdisciplinary communication. Design researchers have already employed various tools such as video, audio and photography. in their capturing processes of designing. The question now is how to organise the large amount of data in a structural way so that the data can be accumulated, analysed, and communicated in an effective manner. To achieve a better management and encourage collaboration of process research, we may consider the development of a hypermedia-based case bank for design studies. The idea of creating and managing a case bank of design studies comes from the AMORE (Advanced Multimedia Organizer for Requirements Elicitation) system developed in the field of requirements engineering for computer system development.

The *AMORE* platform was developed by researchers at the Software Engineering Institute Carnegie Mellon University to support requirements elicitation processes in software systems design (Christel et al. 1993). One of *AMORE's* main objectives is to facilitate "storing requirements in as close to their natural forms as possible to maximize

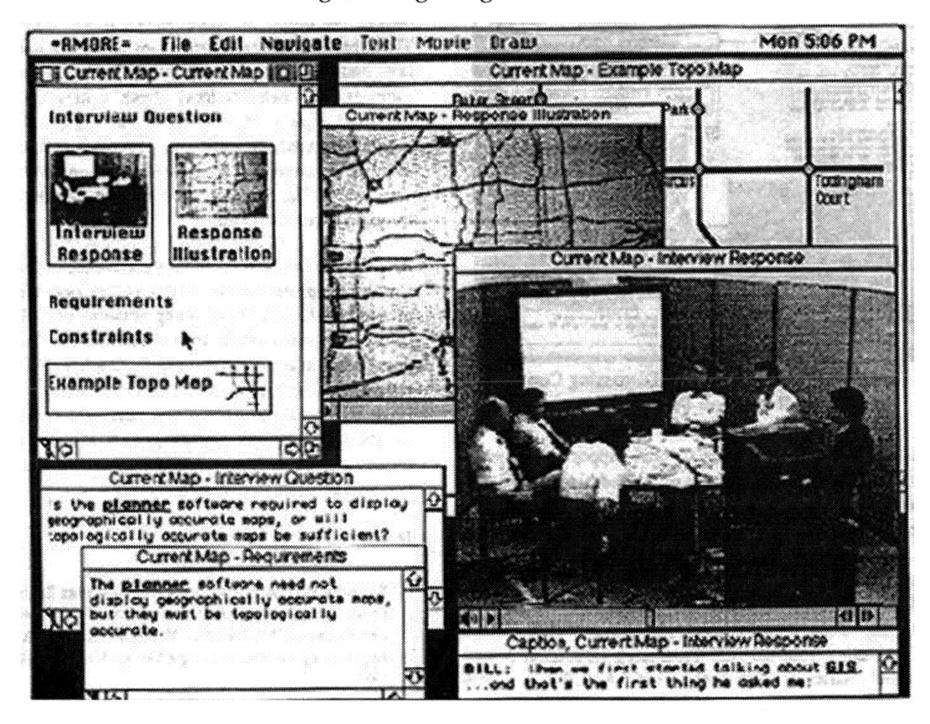

***Figure 8.1** The AMORE information modelling platform developed for requirements elicitation in software system design.*

traceability and to promote understanding of original intentions and motivations" (Wood et al. 1994). The way of preserving the "natural" forms in *AMORE* is to allow raw requirements source material to be recorded in a multimedia database. Requirement elicitors may produce, for instance, interview transcripts in text, video or audio interviews, sketches, tables, and so on. These can all be stored as legitimate multimedia data types in AMORE allowing for structured processing, browsing, and retrieval (Figure 8.1). A multimedia repository can result from using *AMORE* by populating raw requirements data into the database according to the hierarchical or networked organisational structures selected which in turn enable data navigation.

Although the platform was developed to serve the needs in software engineering, a system similar to *AMORE* will be useful for researchers working in design studies. If design researchers use the *AMORE* platform over a considerable length of time, the research may result in a databank containing a large collection of case studies where raw data sets are stored. However, for design research purpose, hypermedia capability instead of multimedia should be employed to make use of hyper-linking in the course of documenting design studies. The diagram presented in Figure 8.2 summarises a general

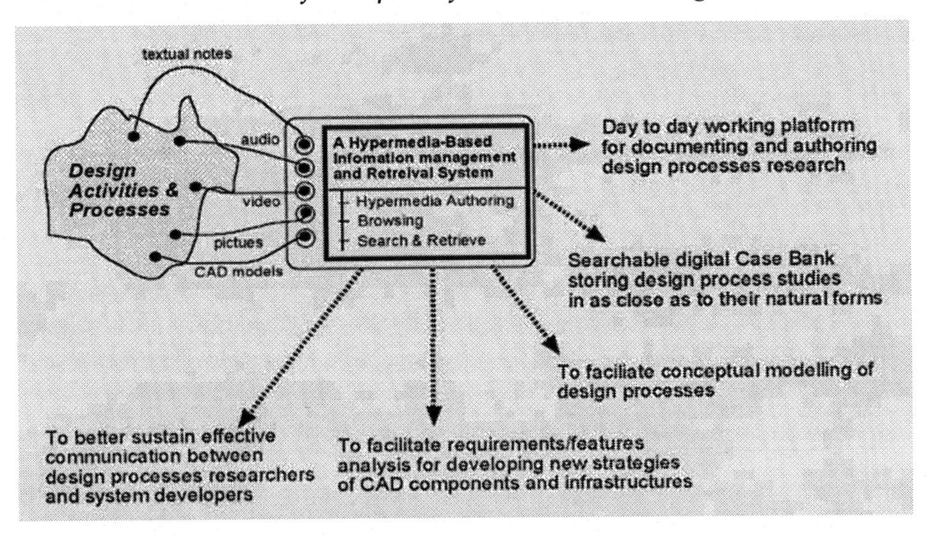

***Figure** 8.2 The objectives of developing a hypermedia-based case bank for design studies.*

framework for developing a case bank for design studies. The aim of the proposed system is to provide a common working resource so that the communication and collaboration of process research and system development can be better sustained. More specifically, the proposed information management and retrieval mechanisms should be developed to serve the following objectives:

Documentation and authoring. Designers often employ a wide range of media when developing designs. Naturally, design processes researchers need to use multimedia tools and documents to preserve the native forms of the research records as much as possible. Process research also requires interpretation of what has been observed from field study. However, it is essential to make the distinction between the design studies as *raw* data collected and the researchers' interpretations of the design activities. Given the same set of field data that records a design activity, it is likely for different researchers to arrive at different interpretations of what constitutes the design. And there is also the need for research findings to be explained by referring to the parts of field data concerned. It is therefore useful to provide facilities for preserving field data in relation to deriving research interpretations.

Searchable data sets. Documents of design studies and process research findings can be more useful if they can be searched and retrieved in as close to as their natural forms. Especially, when a case bank reaches a certain size containing perhaps hundreds of case histories, users of the research resource should be enabled to retrieve data sets effectively. Given the current development of hypertext-based search engines, it is viable to make a large complex case bank searchable.

Conceptual modelling. Constructing models of design processes has been a popular way of communicating research interpretations. By examining a collection of case studies, researchers may arrive at some plausible abstractions that present overviews of key research findings. By including conceptual models of design processes in a case bank, interpretation and abstraction of complex design activity can be traced back more clearly after process research is done. Again, hyper-linking can play a role here in terms of managing relations and referencing.

Requirements analysis. Design studies as documented and authored into the case bank can be dedicated research in its own right. As one of many possible ways of making use of the case resource accumulated, design activity analysis can be conducted to inform the development of new digital media and tools for individual or collaborative design. Given the AMORE environment as an example, we may consider the *coupling* of a requirements or features analysis capability with the case bank on which system research into the development of new components or infrastructures for computer-aided design can be founded.

Interdisciplinary collaboration. In a practice-pulled system development scenario, the collaboration between process researchers and system developers is essential in implementing system functionality in accordance with the requirements analysed. The question is how process researchers and system developers may resolve some sort of common understanding so that the outcome of system development is more or less a response to the requirements identified. The key issue seems to be that findings in design processes need to be translated into kinds of statements or descriptions that system developers can understand and interpret into system constructs and relations. Formal or semi-formal languages of requirements specification may be employed in this regard. However, we should consider that explaining design activities and support requirements in natural languages coupled with direct references to case or scenario studies may well be an alternative to support the interdisciplinary collaboration. And a hypermedia-based case bank can play a significant role in the course of explanation and discussion.

8.2 Project-wide networking and online architectural services

My observations and explanations of collaborative design in the preceding chapters focus exclusively on the communication and coordination among members of professional design teams. As we all know, the entirety of design practice in the real world goes much further than design professionals, involving many other parties in the realisation of building projects. In the second aspect of further research on computer-supported collaborative design, I would like to discuss collaboration involving clients and users. This wider view points to the development of a *project-wide* networking facility that will serve a project team consisting of grouping of professional designers, constructors/manufacturers/suppliers, and clients/users. We may have the networking technologies at hand in constructing project-wide networking, but the impact of such networked communications on the working relations among the various parties involved in planning and executing building projects remains largely unknown. As seen in the late 1990s, there have been tremendous growths in Internet- and World Wide Web-based applications affecting how services in daily life are requested and delivered. It may be timely now for researching into the prospects of Web-enhanced delivering of online architectural services.

Collaborative design with clients/users and constructors

Apart from sound design expertise, success of a building project depends on how well the ABC (Architect-Builder-Client) working relation is maintained throughout the project's lifetime. The process of human communication involved in sustaining ABC relations can range from collaborative project brief development, architects eliciting project requirements from the clients and/or users, architects presenting concept and detail designs, clients querying particular design decisions, etc. As the complexity of project context is ever increasing in the contemporary practice, the demand for good quality of communications between client and architect is getting harder to be handled adequately. The crucial issue is how to achieve efficient communications, that is, being able to identify and resolve misunderstandings or conflicts at the right moments throughout the project's lifetime. A conventional way of dealing with the issue is to have regular face-to-face meetings. Evidence shows that the regularity is either highly constrained by time/human resources or it does not necessarily contribute to better quality of communication.

In his recent book, Andy Pressman has documented vividly many cases of architect-client relations in building projects that took place mainly in the US (Pressman 1995). Not to take the cases as representative to all other societies and cultures, Pressman's study does reveal the potential sources of breakdown in client-architect working relations. And I consider some of the points of breakdown as the starting points for developing project-wide networking support. For example, consider the following passage by Robert Greenstreet as quoted by Pressman regarding legal pitfalls due to poor communication (Pressman 1995, 155):

> Too much focus on the product of the relationship, the building, and not enough focus on the process of getting it built-which includes constant client updating and involvement on all issues-lead to such problems. .... Strong and successful architect-client relationships where legal action is minimized seem to be rooted in solid, well-constructed agreements in which the rights and responsibilities of each side are clearly understood and a 'meeting of minds' has been reached.

Now "constant client update and involvement on all issues" can be substantial workloads for the professional designers to undertake. Also the "meeting of minds" may need to take place many times before the agreements can be firmly reached. The simple question seems to be—Are both clients and architects able to find the times and spaces whenever and wherever meetings are required? The point raised by Greenstreet is that of constancy and timeliness which can be extremely problematic to maintain given the geographical and time constraints imposed. Setting up a project-wide network system in which designers and clients can engage direct or indirect communication without face-to-face meetings seems to be a viable proposition of alleviating the difficulties occurred. Obviously, the potential impacts on the roles and responsibilities of all parties associated with the deployment of such systems need to be better understood.

If clients and architects spend most of their times on deciding what to build, the issues of how to build cannot be dealt with sufficiently without the participation of constructors, manufacturers or suppliers at certain stages of project development. The distributed communications involving builders and suppliers may be even wider in this regard. Richard Saxson, president of the Building Design Partnership, once commented, "Teams need to have face-to-face relationships to achieve the spirit required, but there does not need to be continuous co-location. With physical meetings at intervals, hundreds of suppliers can collaborate remotely to design and provide the elements of a building" (Saxson 1999). The idea of a project-wide networking discussed here is not to replace or eliminate the necessity of synchronous collocated project meetings but to open up further alternative channels of communication such that project participants can be better facilitated to achieve the much desired constant project updating and involvement.

A research programme can be thus defined to explore the deployment of computer networking and groupware that is capable of supporting client-architect communication and collaboration on a building design project. The communication requires project discussion being transmitted via various design mediums (texts, pictures, drawing, verbal conversations, sketches, CAD models etc.) in both synchronous as well as asynchronous manners. The research group led by Paul Richens at the Martin Centre University of Cambridge has recently experimented with an Internet-based communications between the design teams and prospective users of the new Cambridge Computer Laboratory to be built on the university campus (Richens and Trinder 1999). The research project centres on the development and maintenance of an experimental Web site (access limited to the project teams at the time of writing) where the design teams can deposit design proposals in the forms of drawings and 3D virtual reality models and prospective users are invited to view, and discuss and comment on the design proposals. Several technical issues regarding the Web-based technologies employed in the project were identified, but significantly, as reported by Richens and Trinder, the client body has expressed a major concern of how to manage the "growing archive of correspondence, reports, and minutes of discussion which they would like to hold electronically, in the form of cross-referenced, indexed and searchable archive."

Construction of project-wide networking

Since its inception in 1984, early research and development of CSCW systems have mainly focused on facilitating a small size of users to carry out tasks cooperatively. By late 1990s, the scene has changed. We now have the Internet and World Wide Web as the backbone for constructing computing communications and collaboration at much larger scales. The Internet is rapidly becoming the de facto infrastructure for end users to engage in CSCW activities. Facilitated by Internet and the associated networking technologies such as the HTML, VRML, JAVA™ and so on, CSCW developers need not spend time and energy on building dedicated computer networks from scratch. If we consider network-based synchronous and/or asynchronous communications can play a role in the sustaining architect-builder-client relations, we should then look at various types of networking technologies in relation to the Internet when constructing a project-wide network.

Intranet. An intranet can be considered as a mini Internet with a clearly defined network boundary set by a company or an organisation on the basis of the Internet. The

***Figure 8.3** The London Eye Webcam showing a snapshot of the assembly of the Millennium Wheel as progressed on 16 November 1999.*

uses of an intranet are internal to the ownership of the network, which is either invisible or inaccessible to unregistered or unauthorised users. However, an intranet can be constructed with the same set of computing communications protocols as the Internet such as TCP/IP (Transport Control Protocol/Internet Protocol) and HTTP (Hypertext Transfer Protocol). What is required further in order to define an intranet's boundary is a network management regime that imposes certain schemes of digital registration and authentication.

Extranet. An extranet can be considered as an extended intranet that not only serves the users of an organisation internally but also extra users external to the organisation. In commercial extranet applications, the scope of (legitimate) users is extended to registered customers or suppliers. Since the security of data transaction over the Internet has been achieved to a level that wins public confidence, there has been a rapid spreading of extranet-based services on offer. A typical genre of extranet application is online ordering and purchasing systems with which a company can publish the range of products or services and receive customers' ordering and payment, and the customers can send orders and track how the orders are being processed and delivered. Applications of

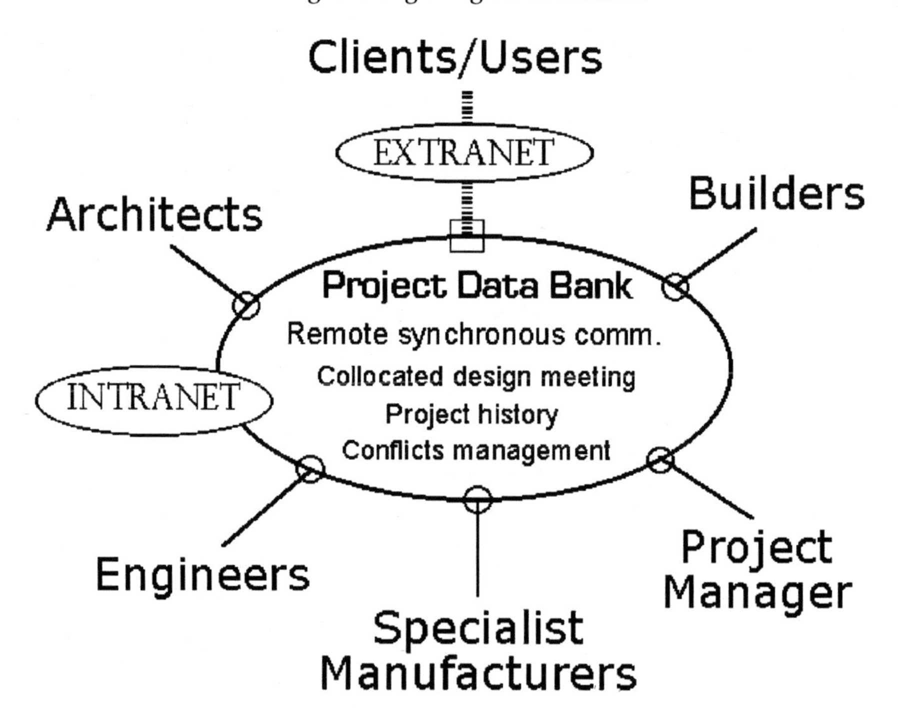

Figure 8.4 Research on the impact of project-oriented networking on project-wide online architectural services.

extranets to facilitate communication in building design projects are beginning to emerge. Jeffrey Huang at the Graduate School of Design, Harvard University, recently conducted a survey of the impact of extranet-based communications on design collaboration (Huang 1999). Huang's survey reveals that the costs of data transaction were reduced and the transparency and accountability of participants' actions were increased due to the uses of project extranets as collaborative media.

Webcam. A Webcam is a digital video camera connected to a Web site that displays computer video images continuously over the Internet.[41] It is getting easier nowadays to come across on the Internet Webcams trained at building sites, which may have been considered by the Web publishers a kind of information and image that the public may feel attracted to. To use the technologies in a professional context, we can think of the deployment of Webcams by architectural design practices as an alternative means to maintain working relations with clients and constructors. To set up this approach, a Webcam-based communication system can be a prominent component of a project-wide extranet. Figure 8.3 shows a snapshot of the WebCam trained at the Millennium Wheel construction site.

Flexibility of project-oriented networking

Design and construction in the built environment is essentially project-based—that is, a team of clients, users, architects, engineers, builders, and so on is formed to work on a particular project, and the team is dissolved soon after the project is completed. The project-wide networking discussed above is to set up a network within a project's lifetime. As soon as the project team dissolves, the project network shall be taken down. Given the project-oriented nature of online architectural services, the technicality and cost effectiveness of running project-wide networking must become acceptable to both the service provider (possibly, project manager or architect) and the users (clients and constructors). From a cost-effective point of view, components of the networking technologies must be reusable across different project situations; and the setting up of a reliable and secured project network should be manageable to the IT personnel of a design firm without calling in specialists frequently. In addition to the potential impacts on the professional working relationships raised earlier, basic research is required to analyse the factors concerning flexibility and user-friendliness of a project-oriented networking technology. Figure 8.4 is intended as a summary of a possible scope of the research into the impacts of project-oriented networking technologies used to deliver project-wide online architectural services.

8.3 Innovative interface technologies for collaborative design

The interface design for better human computer interaction has made great contributions to the user-friendliness of information technology in recent years. In the case of groupware, the interface design for multiple users is one of the most important system development issues in computer supported cooperative work; usability of a system's interface concerns not only a user's interaction with the task-oriented computing applications, but also interacting synchronously or asynchronously with other human co-users. The scope of innovation in interface technologies for collaborative design is potentially very wide. Graphical user interface of a piece of CAD software as often seen on computer screens remains a challenging issue in groupware development. In Chapter 2, we see some examples of multi-user graphical user interface implemented in collaborative drawing environments. A step further, naturally, will be an interface for three-dimensional collaborative modelling systems used by distributed or collocated designers. A question is that if the solutions devised for shared drawing systems are still useable in the case of 3D shared modelling systems. The adding of a further dimension into the shared design space will certainly compound user interface design in achieving effective support for group interaction and for individual autonomy in collaborative design. Another latest trend of inventing new interface technology is the development of *cooperative buildings* that treat the physical spaces we inhabit in as potential interfaces for accessing and manipulating digital information. Following this technological drive, the notion of groupware for collaborative design can be extended into the development of *cooperative workspaces* where design collaboration may take place in a physical space augmented by digital interfaces. I shall discuss more of the emerging interface technologies for collaborative design in the subsections below.

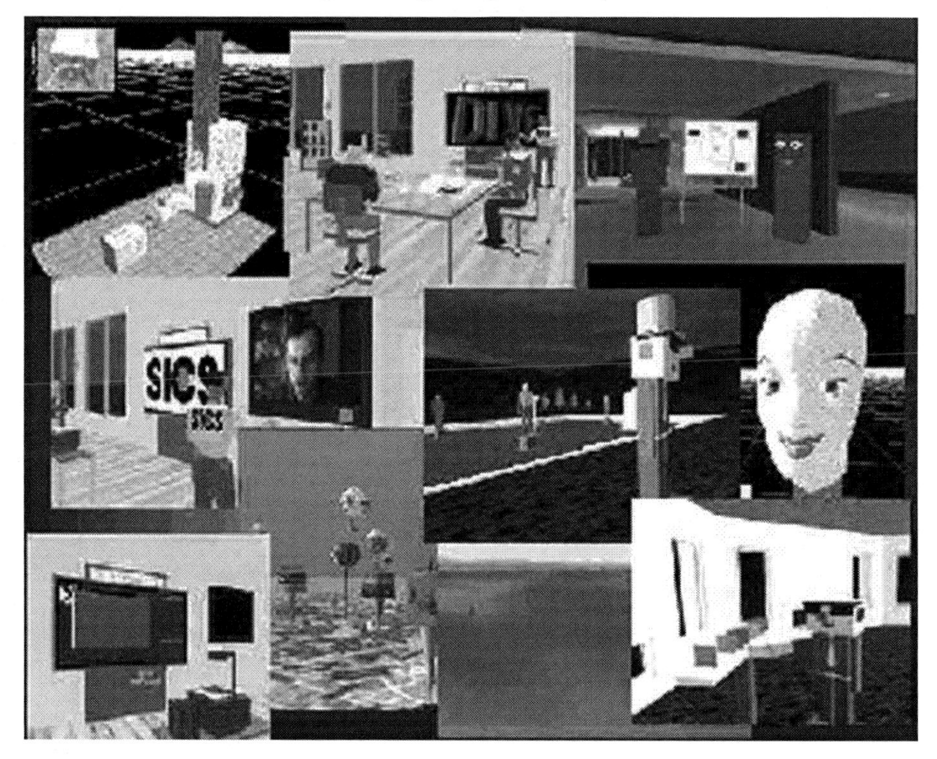

Figure 8.5 The DIVE platform used as a 3D interface for virtual conferencing with avatars representing participants.

Three-dimensional multi-user interface technologies

Having to deal with the implementation of the basic computer networking architectures in the first place in the early developments of collaborative drawing tools, the user interfaces of those tools, not surprisingly, were realised mainly on the two-dimensional computer screens. Although more than one computer screen may have been used in the construction of multi-user real time interfaces, as in the *TeamWorkStation* case for instance, the interface design focussed on how the presence of multiple users should be represented on computer screens. Of course not every cooperative task has to involve three-dimensional viewing and manipulation. But in some design disciplines such as architecture, three-dimensional vision is essential and collaborative design at some stage has to revolve around visualisation in a 3D world. All the knowledge and experiences gained from the experiments of 2D collaborative drawing and writing systems will be valuable to the development of interfaces for 3D collaborative modelling. The question is can the 2D interface designs be extended by simply adding a third dimension, or, is there

Figure 8.6 The projector-and-mirrors system in the Responsive Workbench originally developed by Wolfgang Krüger at the GMD.

a need of looking for fundamentally different approaches? Let us look at two rather recent research systems that addressed the issue of shared 3D visualisation.

Collaborative virtual environments. Employing computer generated 3D models of physical worlds (i.e., virtual worlds) as a new kind of media and user interfaces in group communication has recently resulted in the developments of multi-user virtual environments. A user can navigate through desktop or immersive virtual worlds in various interactive modes such as walk-through. Depositing a 3D virtual world onto a public or private network, a number of users can enter the world and interact with one another synchronously. The Distributed Interactive Virtual Environment (*DIVE*) was developed as a research prototype for constructing 3D synthetic desktop virtual environments and tools which can be accessed by multiple participants remotely for synchronous group communications (Fahlén et al. 1993; Hagsand 1996). The presence and actions of a participant during a live *DIVE* session is represented by an avatar, and through the avatar's viewpoint within the allowed field of vision, the participant will be able to see the presence of other avatars representing other participants in the shared synthetic virtual world (Figure 8.5). Based on the DIVE platform, other distributed virtual environments have been built to experiment with 3D virtual worlds as multi-user interfaces for remote synchronous group interaction. The *MASSIVE*[42] and *COVEN*[43] projects were two such systems that aimed to improve the ability to support richer group awareness and spatial modelling of interaction (see the Web sites for more details).

The examples of experimental collaborative virtual environments have shown us the techniques of adapting desktop-based virtual reality applications into novel interfaces that are capable of supporting group interaction in a 3D space. However, the virtual environments built were mainly user interfaces to some fixed spatial arrangements of interactive events. If we consider applying the technologies in collaborative architectural design, the virtual worlds will be built by design teams as 3D computer models of their

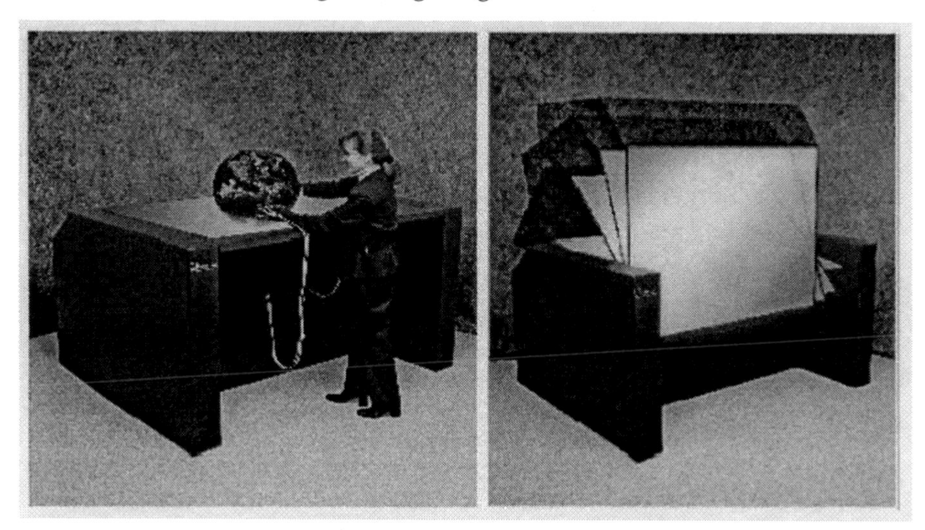

***Figure** 8.7 Fakespace's VersaBench™ system supports collaborative viewing up to six people.*

design proposals. The models are then used as a collaborative virtual environment for the project teams to review the designs developed. Clients, users, builders, and designers taking part in the project can enter the virtual worlds of the proposed building for discussing their concerns of design and construction. It will be interesting to investigate if participants' common understanding of design intents and problems can be enhanced by the ability to see what others are seeing in a virtual world. At their current status, the modelling tools provided by most collaborative virtual environments are rather limited, which cannot deliver the kind of visual sophistication and realism that is normally expected in professional architectural modelling. Perhaps, a way forward is to explore the interoperability between conventional CAD modelling systems and the builders of collaborative virtual environments.

Responsive workbench. Being developed originally by Wolfgang Krüger and his team at the GMD (the German national research center for information technology) in 1993, the Responsive Workbench system was a single-user 3D interactive viewing surface (Krüger et al. 1995; Krüger and Frohlich 1994). Via a projector-and-mirrors system (see the diagram in Figure 8.6), stereoscopic images generated by a computer that sits in the background are projected onto a horizontal tabletop display surface. To view the projected images in 3D effect, the user wears active shutter glasses. A user's interaction with object images projected on the workbench is enabled by a tracking system of six degrees-of-freedom that tracks the user's head, a pair of gloves and a stylus used as a pointer.

By adding a dataglove-based interface for two-handed manipulation and two-viewer tracking (one from each participant), the original workbench environment was later extended by a research team based at Stanford University to support interaction, visualisation and collaboration between two users[44] (Agrawala et al. 1997). Efforts have been made to ensure a perspective-correct view of a shared virtual environment. Two users of the workbench are able to view the same 3D virtual environment when standing near to each other around the workbench and conversing in normal verbal exchanges and gesturing. More research was felt needed to extend the tracking of two-handed manipulation to three or four. Lately, commercial virtual workbench systems were being made available by specialist display technology companies like Fakespace Inc. Figure 8.7 shows a view of the VersaBench™ system marketed by Fakespace, which is capable of supporting collaborative viewing for up to six people wearing light-weight passive stereo glasses.[45]

As seen from the above, in contrast with computer screens as in the *DIVE* project, the workbench interface creates a kind of optical workspace that actually resembles people standing around a table to discuss or make things together. From a 3D visualisation point of view, the workbench approach will appeal to designers as a more direct and intuitive design interface to work with. Depending on further development of interface metaphors, the workbench systems may also appeal to non-professional designers involving in building projects such as clients and users. The ability for clients and users to see and manipulate graphical objects in close correspondence to the 3D real world may thus facilitate better understanding of the design issues and intentions. For collocated synchronous collaboration and meetings, multi-user responsive workbench interfaces can provide useful supports for 3-dimensional visualisation. To professional designers, the usefulness of the interface technology will depend on its integration with other computer-based editing and modelling tools that designers often use.

Cooperative workspaces

A new research forum named as *Cooperative Buildings* emerged lately that aims to explore the integration of information technologies, work organisations and architecture. The first international workshop on Cooperative Buildings was held in Darmstadt in 1998 (CoBuild'98), and the second Workshop (CoBuild'99) in Pittsburgh.[46] One of CoBuild's objectives is to focus on:

> ...ways and means to enable and support flexible forms of communication and collaboration for a variety of groups ranging from small local teams to large distributed organizations, utilizing a seamless integration of physical and digital objects, of real work spaces and virtual information spaces embedded in real architectural environments.
>
> A flexible and dynamic environment that provides cooperative workspaces supporting and augmenting human communication and collaboration.

The cooperative buildings as a whole is an exciting research programme that presents a rich cross-section of research pursued in various fields. Being different from the agenda of *intelligent buildings* in the 1980s, which is more of how to control the environmental aspects of buildings with machine intelligence, cooperative buildings look at innovative ways of turning building fabric into the interfaces between the inhabitants and digital information. Consider "Roomware"—the idea of making parts of a room with its

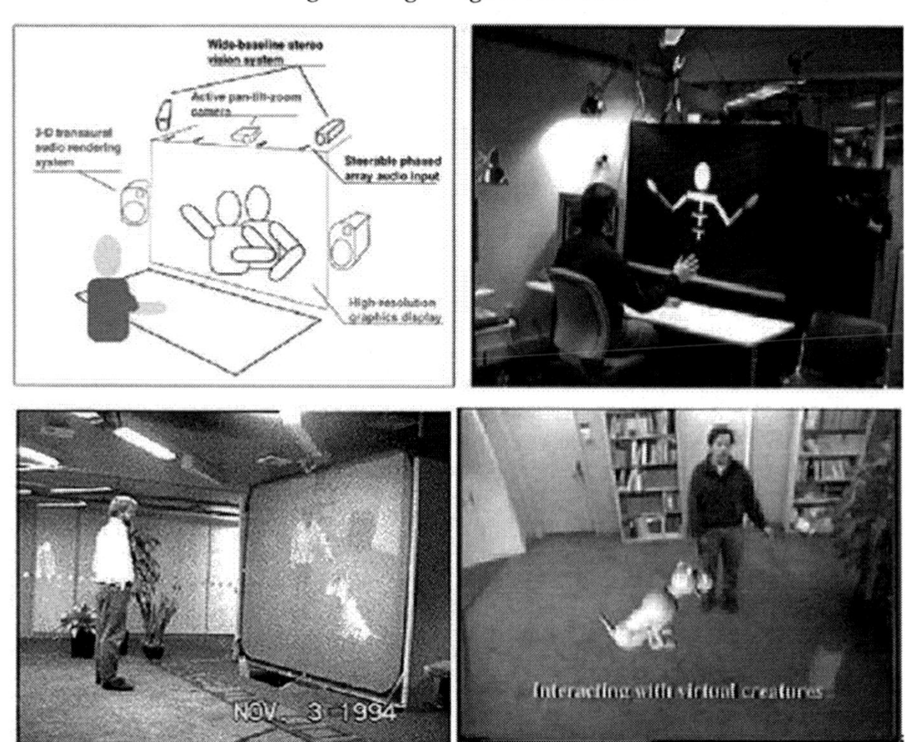

Figure 8.8 *The SmartDesk and SmartRoom systems developed at the Media Lab, MIT.*

windows, walls, furniture, etc. an integral piece of information technology which not only accommodates the needs of dwelling but also the supports for working. The implications of cooperative buildings as interface technologies for supporting collaborative design can be quite radical comparing with those of the workbench or immersive virtual environments. Let us have a closer look at three related research projects in the field of cooperative workspaces.

Smart Desk and Smart Room. The *SmartDesk* and *SmartRoom* projects were experimental computer-based perceptual input and output systems developed by the Perceptual Computing group at the MIT Media Lab. These are devices of combined special hardware and software using cameras, microphones, and biosensors to monitor an individual as he or she works on a desk or in a room (Figure 8.8). The systems were built with a range of underlying techniques drawn from machine intelligence that has the ability to track and interpret human hand gestures and body movements within a physical space (Pentland 1998; Wren et al. 1997). The perceptual intelligence mechanisms mounted onto a desk or room space are then augmented by active agents technology that are able to figure out what assistance may be useful to the person being monitored.

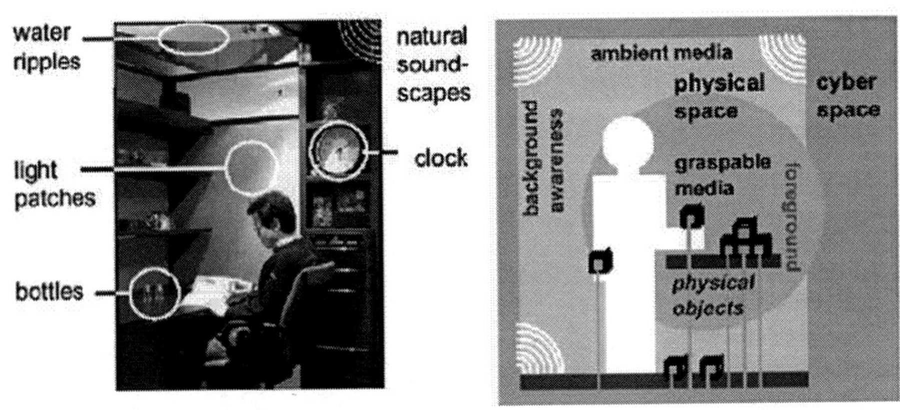

Figure 8.9 The idea and construction of the ambientRoom by the Tangible Media Group at the MIT Media Lab.

Metaphorically, *SmartDesk* was intended to act like an "office assistant;" it "should know your work habits and preferences, remember where you put things, know when you are feeling frustrated or tired, and know enough about your work to anticipate many of your needs".[47] And similarly, a *SmartRoom* will act like an "invisible butler." In their current status, *SmartDesk* and *SmartRoom* are primarily implemented as personal workspace assistive technologies. The point is that the basic science of machine intelligence for interpreting gestures and movements has been developed and the techniques can be further employed in constructing smart workspaces for group working as in the i-LAND project (see below).

ambientRoom. Moving beyond current personal computing interface of looking through small desktop windows, the *ambientRoom* project developed by the Tangible Media Group at the MIT Media Lab attempts to explore the media of "Ambient Displays" as a new form of interface between humans and online digital information (Wisneski et al. 1998). Examples of ambient display media are ambient light, shadow, sound, airflow, water movement in an augmented architectural space (Ishii et al. 1998). To take the instance of airflow, how does airflow serve as an interface for digital information? The research group envisioned that air-conditioners may be computationally enabled to change the flow of air and this change will be perceived by people as conveying some sort of online information via digital communications. The *ambientRoom* was constructed on the basis of the Personal Harbor™, a high-tech individual work suite made by Steelcase Corp., with ambient displays such as light patches representing the number of people sensed in an area, ambient soundscapes transmitting sounds made by people working in another room, etc. (Figure 8.9). Some devices of controlling display media were also developed such as small physical bottles that can be opened or closed as a way of "releasing" digital information into the *ambientRoom*. Experiences from the *ambientROOM* project seem to open up thinking about new possibilities of exploring seamless transitions between background awareness

Figure 8.10 The "roomware" components of DynaWall, InteracTable, and CommChair developed in the i-LAND environment.

and foreground activity. Given that the age of ubiquitous computing is on the horizon, designers may soon work in an augmented workspace in which distributed digital displays and controls are assimilated with parts of furniture and building fabric. Seen from this prospect, future development of groupware for collaborative design should be set free from the flat computer screens of today.

i-LAND. The idea of "roomware" is central to the i-LAND (an interactive landscape for creativity and innovation) project founded by the researchers at the Integrated Publication and Information Systems Institute-GMD. Roomware is defined as "computer-augmented objects in rooms, like furniture, doors, walls, and others" (Streitz et al. 1998). The i-LAND project aims to integrate several roomware components into a combination of real, physical as well as virtual, digital work environments for dynamic (e.g., on demand and ad hoc) teams who work in changing flows of activities. Flexibility and mobility are the key objectives to achieve in i-LAND's roomware which is considered a step forward in comparison with some of the earlier CSCW team room environments such as the CoLab environment seen in Chapter 2. Figure 8.10 shows some snapshots of the current realisation of i-LAND including an interactive electronic wall (*DynaWall*), an interactive table (*InteracTable*), computer-enhanced chairs (*CommChairs*), and a concept called Passage (Streitz et al. 1999). The idea of Passage works around an information carrier called "Passenger" and a receiver "Bridge." It was intended as an intuitive way of transporting data between computers/roomware components and different rooms. The network infrastructure of i-LAND is supported by a combined local area network and RF (Radio Frequency) -based wireless network. The success of the roomware components is to be tested if the integrated real spaces and virtual spaces will allow for dynamic configuration and flexible allocation of resources for project teams.

The number of research projects seen in this section can be considered as examples of innovative interface technologies that are pushing the boundary of groupware for collaborative working. Ranging from the representation of three-dimensional worlds as desktop virtual environments, the projection of 3D models on a multi-user workbench and digital media augmented workspaces, what we see on the horizon is that interface technologies are moving away from conventional computer screens onto more versatile work surfaces and spaces. Indeed, some of them may well be integrated further. For instance, the *InteracTable* in the i-LAND project may well be a responsive workbench, and the *ambientRoom* may have a Smart Desk in it. It is likely that we shall see more experiments in constructing new workspaces that can respond to changes in work organisation and to inventions in building or furniture making that open up opportunities of integration with novel digital interfaces. Research and development of groupware for interfacing design collaboration can therefore be extended to the augmentation of design offices into cooperative workspaces highly adaptive to project-oriented networking that is becoming evermore dynamic in the field of building design and construction.

8.4 Software agents as parts of a project team

The further research on innovative interfaces technologies discussed above is mainly for developing augmented or virtual common workspaces with which project participants can create and share visualisation of design intents and actions. There is an equally if not more important aspect of further development in groupware for collaborative design: managing design knowledge or, to put it formally, knowledge management in design collaboration. This is not intended as a follow-up of the split perceived between form and function (or, performance) in contemporary architecture making. However, for the purpose of developing computing system components, it may be necessary to focus on knowledge management as separate issues from those of 3D visualisation. In practice, as the EdCAAD research group has tried to address in the mid-1980s, system development as a whole should provide designers with a modelling environment that can handle integrated knowledge and graphics. As a sequel to the further research on innovative collaborative interfaces, I shall in this section discuss a list of topics for developing groupware components in support of knowledge management in design collaboration. Representing and managing knowledge as *agents* has become one of the hottest topics in recent artificial intelligence research. At various points of discussing the bottom-up and top-down teamwork patterns, some of the group communication supporting mechanisms were indeed referred to as agents. Given that the current agent technologies are moving rapidly to embrace networking and mobility, it seems appropriate to suggest further research into the prospects of intelligent agents, or, more appropriately, software agents, as parts of a project team.

A software agent is basically a piece of software with built-in intelligence to perform some delegated tasks. The nowadays agent-based technologies may be traced back to the programmes of Expert Systems or Knowledge Base Systems developed by people working in the field of artificial intelligence in the early 1980s. As briefly mentioned in Chapter 1, an expert system is a collection of facts (data) that can be processed by the rules and machine reasoning mechanisms encoded in the system. Expert or knowledge base systems have resulted from the basic research into how knowledge could be represented and manipulated in computational systems, but the applications of such systems have been highly domain specific. This may explain why so far in the field of building design, developments and applications of knowledge base systems weighed more on the side of building performance analyses than design syntheses. Early appearances of *agents* can be seen in the development of *Distributed Artificial Intelligence* (DAI) systems. A DAI system aims to solve a specific problem via a network of intelligent agents each of which is capable of handling a particular aspect of the problem. The ACDS (automated-configuration design service) system developed at the University of Michigan is an example of DAI system, in which various types of agents dealing with user interface, design constraints and specification-solution matching were combined to produce optimal configuration design in mechanical-electronic systems (Birmingham et al. 1993). Concepts of computational agents continued to evolve in the AI community and they are increasingly construed as service-providing human agents as in, for instance, travel or estate agents. Pattie Maes, founder and director of the Agents Group at MIT's Media Lab, once outlined what she meant by agents as follows (Shneiderman and Maes 1997):

> ... a new approach to user software, a new way of thinking about software that end-users have to use. In particular, the way in which agents differ from the software we use today is that a software agent is personalized. A software agent knows the individual user's habits, preferences, and interests.

Following Maes' definition of agents, the paradigm of agent-based system research seems to be shifting from differentiation and distribution of problem-solving capabilities to user delegation and personalisation. A newer generation of software agents is expected to differ from conventional software in that they can be (semi-) autonomous, proactive, and adaptive. In the following subsections, I shall discuss the various roles that can be explored in the development of agent-based technologies for facilitating knowledge management in collaborative design.

Agents for retrieving product information

Building designers or specifiers often need to consult volumes of building product information before reaching decisions about what materials and specific building components to use in a particular project. Suppliers or manufacturers usually supply product information in the form of product catalogues. From a designer's point of view, quantities, dimensions, and performance standards are perhaps the key aspects of information to be specified. A building component supplier, on the other hand, may catalogue the range of their products and, more importantly, what performance standards their products conform to. To produce a satisfactory building design specification is to achieve best possible matching between design requirements and the availability of building products conforming to certain standards of performance or qualities. Given the complexity of modern constructions, to identify and track all the matching in a building project are no trivial tasks. Attempts have been made recently to devise ways of making building product information digital and online. In the following, we shall look at two research projects of a similar nature exploring the potentials of interconnecting online digital product data supply with conventional CAD systems.

The PLAid (Product Library Assistant-an intranet for designers) prototypical system developed by Richard Coyne's research group at the University of Edinburgh is intended mainly as an organiser for designers to manage design project database in connection with online product information on the Internet (Coyne et al. 1998; Coyne and Lee 1997). As more and more suppliers and manufacturers of building components are publishing the information of their products as Web pages written in HTML, building design practitioners need to find a way for creating and sharing project directories that hold structured data nodes pointing to the Web pages of the products chosen. In PLAid, designers are able to pick up elements represented as graphical icons from project directories and examine product descriptions and specifications that are made available at the linked Web sites. PLAid also explores the benefits of treating Web pages' linkages as attributes to be held in project libraries supported by a conventional CAD system so that CAD users can get access to Web-based product information directly without leaving the system. In doing so, a CAD system is made network-aware. The research team envisages that in the foreseeable future all building product information will be made online but efficient searching and filtering mechanisms are essential in dealing with a vast quantity of distributed information resources.

The ARROW (Advanced Reusable Reliable Objects Warehouse) project was a UK initiative directed by the IT research group at the Building Research Establishment (BRE) to develop a prototype of Internet-based *virtual warehouse* (Amor and Newnham 1999).

The warehouse system aims to provide information services to building designers/specifiers and building product manufacturer/suppliers. By developing an Internet-based data modelling framework, the project has demonstrated that it is possible to identify any of a range of manufactured products meeting specific design criteria. This is done by designers interacting with the system's knowledge-based query handler and specifying the parameters of product types related to their design at hand. By matching the parameters specified, the system's central search engine is able to return all products recorded in the warehouse that satisfy the requirements. The searched product information is then delivered in a form operable by CAD systems and other performance analysis tools. A precondition for the ARROW methodology to work efficiently in a wider context is that manufacturers and suppliers are willing to publish electronic information of their products in some publicly agreed data models or standards. As pointed out by Amor and Newnham, the chief architects of ARROW, the acquiring of structured data from all manufacturers and suppliers is the biggest challenge facing those who would like to develop ARROW-like systems. They suggested that further development of toolkits for manufacturers and suppliers to ease the cost of product data provision will be a major facilitator in the development of ARROW-like systems.

Agents for negotiation in collaborative design

In the top-down scenario of team working I described the need for negotiation among design participants to resolve conflicts in making design changes. This is considered much a human communication process in which designers discuss implications of design changes and explore alternative ways of developing domain designs. In perhaps some rather limited areas of building design, design preferences or constraints (such as geometry or performance) can be represented as agents that can act upon design changes proposed by team members. The authored agents may carry out routine checking and generate responses to proposed changes in accordance with what is acceptable and what is not as laid down by their designers. One of the potential benefits of doing so is to achieve a higher degree of *concurrency* in areas that allow for such computational treatments. Concurrent design may be necessarily involved at some stage of design development where an overall optimal solution cannot be reached without multiple sub-decisions being made at the same time. In the absence of some project members, concurrency may be maintained if designers working in different domains can delegate some parts of their design knowledge and judgements into some form of software agents that are deployable on a project-wide network.

A software agent can act on messages sent by human designers or other domain agents in real time by processing the requests or proposals contained in the messages and sending replies on the basis of its knowledge base. As the designers return to their workspaces, their software agents will prompt with a record of what has been negotiated with whom, along with what knowledge has been applied. Obviously, there will be many areas of collaborative design that cannot be subject to agent-mediated negotiation where stable bodies of design knowledge are simply non-extant. Agents for negotiation in collaborative design may be more applicable if designers have identified during project development areas of communication that are stable but mundane and tedious.

Software agents for negotiation attracted much attention in the recent emergence of e-commerce. Automated negotiation has been identified as one of promising areas in developing Internet-based trading by introducing agent technologies (Nwana et al. 1998; Sandholm 1999). While the business aspects of agent-based buying and selling may be not so directly relevant to negotiation in design collaboration, I believe that there are some useful lessons to be learned from the infrastructural aspects of implementing multi-agent negotiation as pioneered in the *Concordia* platform (Koblick 1999).

Agents for managing project history

The lives of building design projects can vary considerably from project to project. Some may only last for a couple of months, while others for years till completion. Information and knowledge about the history of how a project has evolved can be valuable in managing larger scale projects. Especially, projects involving a high turnover of design personnel can present considerable difficulties for members to collaborate if they take part in the project at different stages of project development. By getting access to some records of project history, late or rejoining design participants can keep themselves updated so that they may be helped to tune into the latest status and pace of team working relatively easily. What we see in design practice is that project history is normally held in human memories supplemented by perhaps a huge amount of design drawings, contractual documents, meeting minutes, correspondences, and so on. Designers do not normally keep explicit records of project development history. If knowledge about project history is considered desirable by a project team, computer-based mechanisms and processes for recording design events are worth deploying. Agent-based technologies may play a role in this respect. Software agents can be either passive or semi-autonomous depending on project requirements. The aim is to facilitate the generation of longer-term project memory that can be queried by project participants as time goes by.

We can even imagine a multi-agent approach in which each agent is capable of logging significant events as seen from a domain-specific viewpoint. The resultant multiple records may later converge into a single coherent account of project history. Further research may focus on how project history can be presented in ways pertinent to participants' queries; software agents can act as intelligent interfaces for searching and retrieving particular sessions of events together with links to associated design drawings and documents that an end-user enquires. CSCW researchers working in the area of organisational memory have produced system-oriented and field studies regarding the formation and uses of social memories associated with work groups (see, for instance, the work done by Mark Ackerman and others on the *AnswerGarden* system (Ackerman 1998; Ackerman and Malone 1990), and the theory and tools developed by Gerhard Fischer's group to support long-term collaborative design (Fischer et al. 1999). However, collaborative design may have a different emphasis on the nature of social memory that tends to be project-oriented rather than organisation-oriented. Building design project teams can be considered as temporary organisations which can be short-lived as the projects soon come to an end.

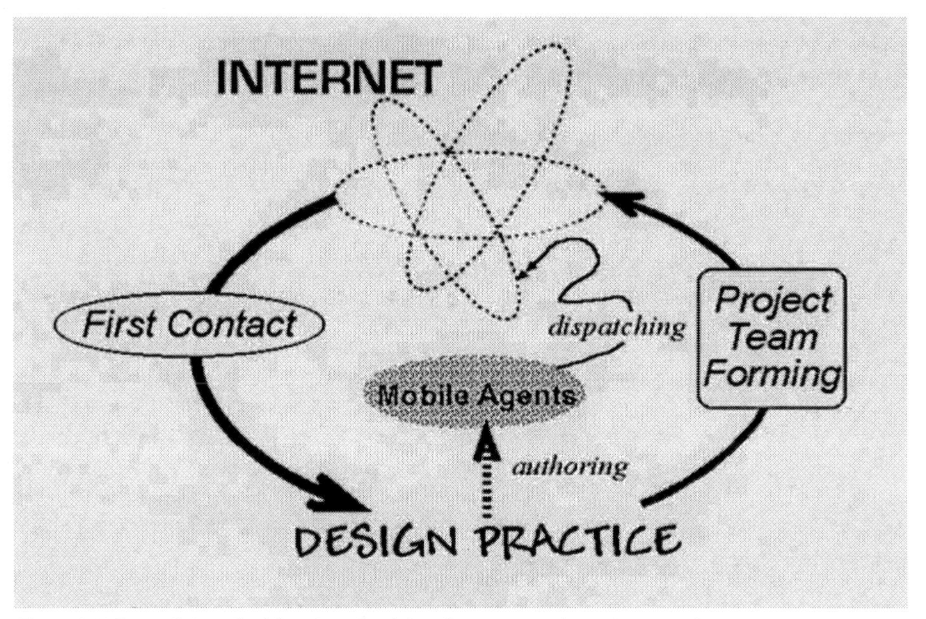

Figure 8.11 Research into the delegation of mobile software agents for project team forming.

Agents for making first contacts

Personal connections and relations seem to be the major factor that governs how building design project teams are formed in practice. We often come across well-known cases of design partnerships that accomplish a significant number of projects through collaboration lasting for many years or decades. Less established design practices may not be in a position to rely on readily formed project partnerships. Or, occasionally, well-established practices may need to work with different specialist groups due to unusual project requirements that can arise. In client-driven projects, project managers may be employed in the beginning to put up project teams according to a set of criteria. In more open-ended scenarios, it may take much less straightforward routes for partners to find one another in forming a project team. Given that the geographical distance is increasingly less a barrier for people to know and work with each other, the process of forming project teams may become an actual part of collaborative design. However, increased chances of team forming do not necessarily mean that it is less time-consuming to initiate successful project partnerships. The problem seems to be how distributed design groups who may know little about one another initially are able to make informed first contacts with a view of forming a project team.

As a potential component of next-generation groupware for collaborative design, agent-based technologies may be employed by design individuals or groups to establish first contacts through the Internet as an alternative way of initiating team building. The kind of software agents envisaged will not do much with buying and selling through the

***Figure 8.12** An example of group work in design education: the Sheffield Urban Study project. (Photograph by Peter Lathey 1998, School of Architecture, University of Sheffield)*

Internet as in e-commerce. Instead, individuals or work groups will be enabled to construct their own software agents and then dispatch them into the cyberspace to actively search for project partnerships. The behaviours of the software agents will be governed by perhaps an Internet-based public forum and interaction protocols as well as the packages or statements specified by their authors which can contain instructions and messages regarding, for example, cooperative design expertise sought, project initiatives interested, or even agreeable working ethics, etc. By judging the information collected and processed by the contact-making agents, the individual or group authors can then decide if further direct contacts are desirable. Future research on groupware for collaborative design may set out to explore the feasibility and prospects of agent-mediated team building processes. Among the most relevant technologies to look out for will be Internet-based distributed computing communications. Newly available computing platforms and languages such as JAVA™ and XML are driving the developments of *mobile agents* technology into a new paradigm of distributed computing that takes into account Internet-based communications (Glushko et al. 1999; Lange and Oshima 1999; Wong et al. 1999). Many other related issues should also be considered such as end-user authoring of mobile agents, the setting up of public virtual forums and communications protocols for dispatched agents to meet and exchange information, end-users' interaction with mobile agents in retrieving collected information, etc. (Figure 8.11)

8.5 Groupware technologies and design education

At the dawn of the twenty-first century, we are beginning to see the impacts of computing communications and network technologies on all types of knowledge-based service industries as well as aspects of people's daily life. The activity and enterprise of architectural/urban design and construction for the built environment will be part of the evolution if not always perceived as forefront runners. The general public as a whole are demanding design products of more holistic properties that better meet up with complex issues not confronted before such as environmental sustainability and social responsibility. On the other hand, current education and training of future design professionals seems to have difficulties in not continuing the route of division and specialisation. To somehow reconcile the two trends, design education and training should consider the needs for fostering students' knowledge, skills, and experiences of working in groups consisting of member students from perhaps different departments. If collaborative design learning is to be strengthened and enriched on the basis of the current curriculum, groupware technologies can be important ingredients in forming more forward-looking education programmes.

Teaching and learning in groups are by no means new to schools of architecture or engineering. Taking the School of Architecture, University of Sheffield as an example, the Sheffield Urban Study project, conducted during the 1998 academic year, involved a large group of diploma level architecture students who were organised into smaller teams to undertake a large-scale study of urban history. By constructing jointly a physical model showing the core of Sheffield City at the time around 1900, the group work has produced a valuable collection of data for future uses and revealed a city unseen before (Blundell Jones et al. 1999) (Figure 8.12). There are other collaborative projects of interdisciplinary nature often involving architecture and structural engineering students working together as project teams. Seeing this as an example, I think we may make the point that there seem no shortage of education programmes designed to implement the idea of single or multidisciplinary group working. The question now is how elements of collaborative design computing may be integrated into these programmes so that students can acquire knowledge and skills of collaborative working in networked digital media. There are basic concepts about collaborative computing to be taught, and new types of teaching programmes can be developed based on creative applications of digital media links.

Learning basic concepts and skills of collaborative computing

Supporting design collaboration has not been a major feature of conventional computer-aided design packages that are still commonly used as CAD teaching packages at most schools. Major CAD developers such as MicroStation and ArchiCAD are beginning to incorporate limited teamwork supporting features in recent software releases. Some of the collaborative features provided in the current system packages may not be appropriate from a design education point of view as they are implemented to reflect certain models of document management or organisational hierarchy that are not necessarily true in changing contexts of design practice. For design students to acquire more basic concepts and skills of collaborative computing, we may need to consider a wider range of choices in defining proper contents of teaching programmes.

Collaborative writing and drawing. A good starting point for students at a beginning level may be structured around collaborative writing/drawing sessions where students can experience what it is like in creating shared information space in networked digital media and what the potential problems are to conduct collaborative working under various spatial-temporal patterns (i.e., collocation vs. remote, synchronous vs. asynchronous). Using general collaborative tools like *Dolphin*[48] or *Habanero*[49] (Figure 8.13), students in groups can learn to how to create and review their own documents in remote synchronous sessions.

Hypermedia authoring. Hypertext authoring has recently become more widely practised by non-experts with the availability of HTML editors such as Microsoft's FrontPage or Adobe's DreamWeaver, which were developed mainly for creating Web-based documents. These HTML editors provide graphical user-friendly interfaces for authoring and editing HTML document in a way similar to those of ordinary word processors, saving authors from direct coding in the HTML language. More importantly, the hyper-linking mechanism can be employed to produce hypermedia documents containing data on multimedia formats. And the method of hyper-linking actually provides a generic mechanism for facilitating group work. Adopting the hypermedia authoring approach, students can be organised to collaborate on, for example, modelling an urban context (as in the Sheffield Urban Study project mentioned earlier) or a project site by generating hyper-linked contextual literature and 3D models that can be later used as shared references for developing urban or building design projects (Peng and Blundell Jones 1999).

Collaborative 3D modelling. Generally speaking, constructing 3D models of building spaces on conventional CAD platforms is highly technical. Extending the current single-user 3D modelling platform to allow for multi-user collaborative modelling may not be the right route to follow. Although responsive workbenches have shown some promising results for collaborative 3D modelling, it seems not viable for a university department to deploy a large number of the workbenches even if a specialist workshop is set up. A more appropriate approach in my view will be a VRML-based platform. Being similar to HTML, VRML models are three-dimensional graphical models of which any parts can be authored to point to other documents of various formats via hyper-links. The doorway of a house modelled in VRML, for instance, can be linked to a 2D detailed elevation or sectional drawing of the doorway recorded in a separate file. With VRML modelling, students can achieve not only 3D visualisation of buildings and spaces but also 3D graphical indexes to a potentially vast amount of project information. Again, hyperlinks can be used to facilitate teamwork in 3D modelling. Some industrial VRML developers have made available software components for viewing 3D VRML models deposited on the Internet or an intranet. Often implemented as plug-ins for Internet browsers such as Microsoft Internet Explorer or Netscape Navigator, VRML viewers like blaxxun Contact[50] by blaxxun interactive or Viscape by SUPERSCAPE[51] are capable of supporting multi-user interactivity and navigation into 3D designs. This facility may be used to promote students viewing their 3D designs collaboratively as though they are

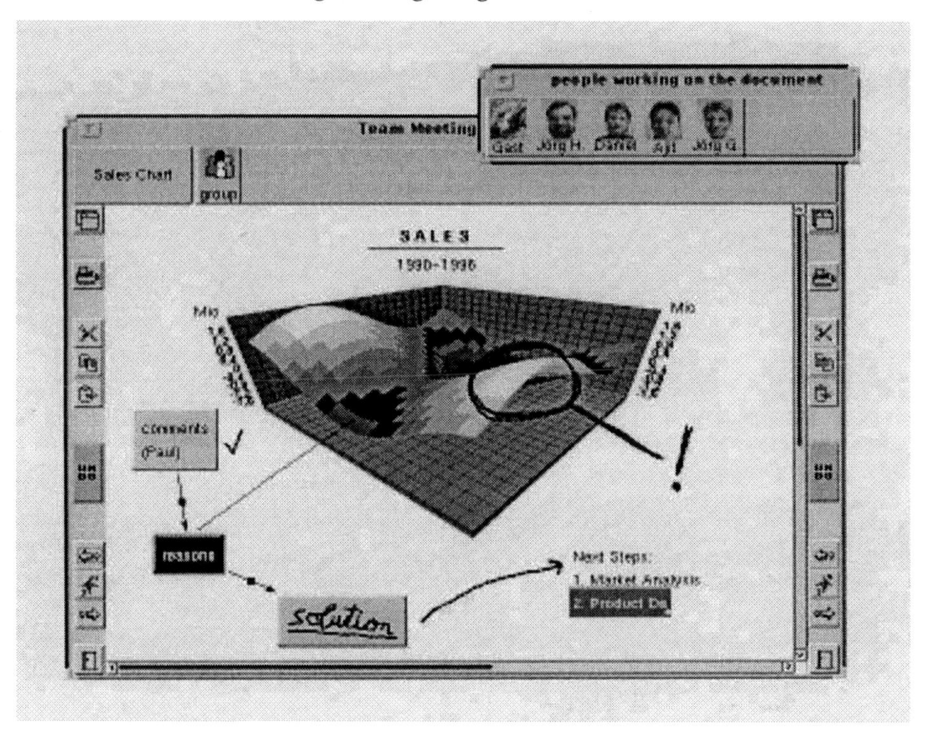

Figure 8.13 (above and right) *Using Dolphin and Habanero as general purposed platforms for basic collaborative computing.*

walking together around the spaces they have created jointly. Furthermore, design tutors may also deliver project tutorials by entering the virtual worlds together with the students and discussing the problems of their 3D designs.

Innovative applications of digital media links

Once the basic concepts and skills of collaborative computing are familiarised, students will be in a better position to take part in networked learning at a larger scale. The focus of learning will then be the skills and experiences of engaging in different modes of design communication via digital media links. Assuming that there are no practical problems of acquiring the human and system resources required, consider the following options as examples of how digital media links may be applied innovatively in design education programmes that will take networked design learning beyond the boundary of a single institution.

Virtual design studio. Unless taking part in some exchange programmes, design students seldom have the opportunities of undertaking learning programmes beyond

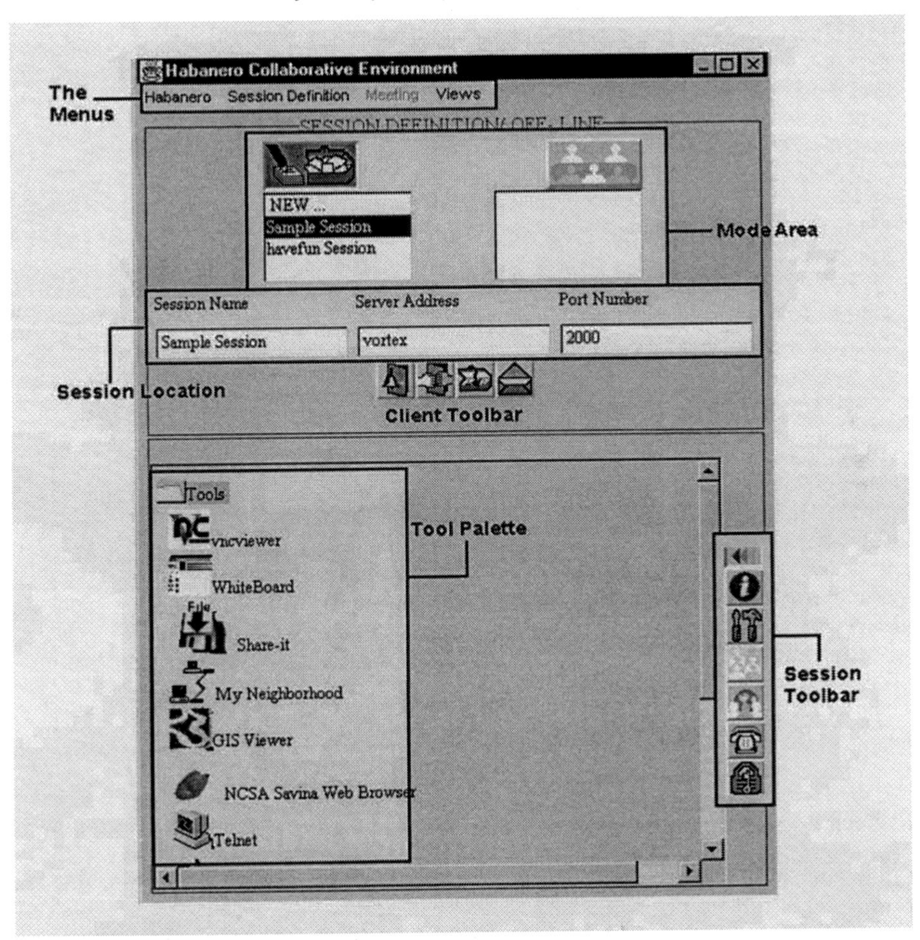

their own registered institutes. Physically moving a group of students from one place to another for joining another group for developing collaborative design projects will demand substantial logistical resources for the organising parties even for a few days. However, the experiences of learning and working in different social cultural milieus are highly valuable from a design education point of view. After all, design is about a journey through which the meeting points of technologies and cultures are found. The idea of virtual design studio is to create such a learning environment in which students at different parts of the world are enabled to collaborate on design projects. In fact, experiments of virtual design studios have been organised and reported by various academic institutes around the globe (see, for instance, Chen et al. 1994; Chiu 1995; Vásquez de Velasco and Trigo 1997 among others). In these experimental virtual studios we can see examples of how long-distance studios may be conducted and what asynchronous or synchronous collaborative techniques can be employed in planning further experiments. In many ways, the virtual design studio approach may prepare students for their future collaborations with design firms located in perhaps distant countries.

Virtual design review. Formal design review or critique is often considered as an essential part of project work in architectural design education. Students are expected to conduct verbal presentations on the basis of the drawings or models produced. To many, this is a communication process simulating how designers communicate design intentions with their future clients and users. On the occasions of major or final design critiques, studio-based teaching staff often invite well-known practitioners or scholars to form a project critic panel. If virtual design studios can be set up among different schools for long distance collaboration, it seems that *virtual design review* should follow naturally. This will typically involve the setting up of video and audio links among participating sites, supporting real time virtual conferencing. As a part of the Design Studio of the Future initiative at the MIT, experimental virtual design reviews have been conducted between the MIT and the Cornell campus (Shelden et al. 1995). In the UK, a similar initiative of virtual design review was held between the Mackintosh School of Arts and the University of Strathclyde (Lindsay and Grant 1997). Both projects have reported problems of matching traditional design review practice with the communications technologies. As the cost and technical barriers of virtual conferencing are lowering, remote synchronous design reviews involving geographically distributed groups will gradually become a more common practice. As a preparation for the not too distant future, it is important for students to be exposed to virtual conferencing experiences which may present them a different set of communication issues and skills to be aware of.

Linking lecture theatres with construction sites. Can design teaching and learning processes in an educational setting be connected with building events in the real world as online resources? Most design educators would probably agree that experiences of live construction processes could be beneficial to design learning processes. But organising field trips to building construction sites is either impossible or not cost-effective. The proposed research aims to conduct a pilot scheme in which multi-media digital links are established between lecture theatre and construction sites or specialist workshops. Several goals are to be achieved via the digital links: (a) two-way real time communication between learners located in the lecture room and instructors/builders at the construction site so that questions can be asked and answered; (b) generation of digital media archives as learning resources in building construction allowing for playback; (c) generation of digital video documents as construction research data allowing for video analysis. And we should also start to explore digital media links to construction sites in the real world so that students can engage in two-way communications with people on the building site, which may be thousands of miles away from the lecture theatre. And the system used can also generate digital archives for later uses in learning or research on construction. More research is required to look into the opportunities offered by digital media links between building construction sites, manufacturing workshops and the lecturing theatres. Design educators and design students alike now face the situations of being seemingly burdened by the ever changing pace and complexity that comes along with emerging design technologies. There are opportunities of developing new educational programmes by introducing innovative uses of digital technologies. However, there are also the issues of losing traditional architectural sensitivity and authenticity due to the crudity of present technologies or improper uses of them. There is an urgent need to develop alternative pedagogical frameworks and teaching materials that will enable a better matching between design education objectives and learning of particular systems of digital tools and media.

Summary

It is expected that future computer-aided design will be developed essentially as groupware that will not only support individual design processes but also facilitate interdisciplinary collaborative processes via advanced computing communications. Several aspects of further research into the development of CAD systems as groupware are discussed in this end chapter. Firstly, there is the need to continue to enhance basic (practice-pulled) research on how collaborative design in the field can be better understood so that the requirements for system support can be elicited and described more adequately. We discuss the idea of a hypermedia case bank for design studies that will enable researchers to store and organise field data in their natural forms, and the resultant databanks can be used by researchers involved in different fields of design studies for collaborative research purposes. Secondly, as the reliability and affordability of intranet and extranet based technologies continue to advance, it is now practical for design practitioners to deploy project-wide networking and make their design services online to their clients, users, suppliers and constructors throughout a project's lifetime (i.e., from project briefing to building commissioning). However, research is required to investigate the potential impacts of the networking technologies and online communications on professional issues and designer-client-builder relations.

Looking at some of the most recent developments in digital interaction, thirdly, we envisage that innovative interface technologies for collaborative design are forthcoming. Conventional computer screen based CAD environments need to be extended with richer and more versatile capabilities of three-dimensional interactions that can be integrated more fully with the physical workspaces designers occupy no matter however dynamically. Fourthly, speculating about how a project team might be formed in the first place, we consider the possibility for design practitioners to author and dispatch software agents into the Internet and make first contacts with potential partnerships that match their searching criteria. Agent technologies can also become a valuable component of an innovative CAD environment if intelligent agents can actively search for satisfactory building products published on the Internet and advise designers/specifiers the options available in a timely manner. Finally, while computing communications is fast entering design professions as a mainstream enabling technology, teaching and learning of CAD skills at most architectural and engineering schools remain oriented to largely individual processes. There are immediate needs to develop new education and training programmes that will make creative uses of groupware technologies a part of the learning processes. Design educators should also consider the potentials of deploying groupware to facilitate teaching at academic institutions connected in real time with design and construction events in the real world.

Glossary

Below are the terms and notations used in this book (Chapters 3 to 6, mainly) for describing *constraints on collaborative design* in a shorthand form. Only general readings of these terms and notations are given; a system of formal semantics is *not* intended here.

Terms in the bottom-up scenario:

Individual Modelling Spaces (IMS). Workspaces created and evolved by designers individually for modelling design expressions (e.g., diagrams, drawings, or any other graphical/textual constructions) targeted at a particular design aspect or domain of a design project.

Group Modelling Space (GMS). A common workspace created and evolved by members of a design team jointly for modelling the integration of design parts contributed by each member into larger design wholes.

Individual Object World. A representation scheme formed by a designer's abstraction act performed (individually) in his or her own IMS.

Local Design Expressions (LDE). Instances of *LDE* are design expressions showing, for example, what spatial forms or particular functions are modelled by the designer when embarking on domain-specific design tasks. We use LDE_a to denote a local design decision made by a designer named *a*.

Shared Integration Schema. Design concepts or methods that are developed jointly by members of a design team for the purpose of performing the integration of individual contributions into design wholes. A collection of such shared concepts or methods can be correlated into an *integration schema* that can be constantly employed by the design team to handle design integration at a larger scale.

Common Images (CI). Common images are pictorial representations that are generated by design members putting proposed *LDEs* together and then transforming them into single compositions under the operation of shared integration concepts and methods.

Common Design Metaphors. Common notions or images reached jointly by members of design group that are taken as common design references to either *what a common image looks like* or *how a common image functions* during design communication.

Domain Design Tasks. An individual's interpretation of a common image may bring out some significance or role of the individual's work in the context of an emerging whole. A domain-specific design task is formed in an IMS as a result of an individual's interpretation of *CI*, which may contain items of information useful in guiding further domain design development.

Terms in the top-down scenario:

The terms **Group Modelling Space, Individual Modelling Spaces,** and **Individual Object World** appearing in the top-down pattern share the same readings as they are defined in the bottom-up pattern.

Shared Construction Set (SCS). A set of modelling primitives introduced by members of a design team for the modelling of **Common Generic Structures** (see below) in a group modelling space.

Common Generic Structures (CGS). Common generic structures are 2D or 3D objects, representing, mainly, a kind of spatial framework or skeleton that is constructed and can be used by all participants working in different domains of a design project. There are several important properties observed in common generic structures, including *deformability, multi-perspectiveness,* and *genericity.*

Derivative Structures. Derivative structures are 2D or 3D objects that are produced by applying derivative operations onto a state of a common generic structure. Images of derivative structures, once imported into individual modelling spaces, can serve the individuals as design referents or outlines in generating more detailed designs of particular design aspects.

Domain Design Expressions (DDE). Domain design expressions are the outcomes from designers substantiating derivative structures with design expressions that contain specifications of domain-specific design substances or properties. Domain design expressions are slightly different from the local design expression in the bottom-up pattern. Instances of DDE are designers' constructions of domain designs on the basis of the derivative structures; while instances of LDE are more of the outcomes from designers' free design intentions.

Local Design Assessment. An individual gives an interpretation (or an evaluation) of domain design expressions arrived at an IMS. Since each domain design expression is generated in relation to what derivative structures are underlay, the resultant DDE, when viewed and judged by its creator from a domain-specific design perspective, is a *design consequence* of a state of a common generic structure.

Symbols for denoting Constraints on Collaboration

Δ (State Change). Any state change occurring to a type of design state. For instance, $DLDE_{john}$ refers to a state change in a local design decision made by a Designer named *john*.

E_1, E_2, E_3, ... (Distributed Entities). A number of representation entities that are distributed over several individual modelling spaces. For instance, LDE_a, LDE_b, ... refers to a distribution of information entities consisting of a local design expression made by $Designer_a$, $Designer_b$, and others.

Σ (Collective Presentation or Construction). States of, for instance, multiple local design expressions or changes according to local design decisions are collectively presented (assembled) by participants in a common visual space during a design meeting. S is also used to denote a resultant common set of constructs developed by collaborative designers.

Glossary

$\rightarrow$ (Leading To). A symbol denoting the *flow of information* (in a situation-theoretical sense) from what is contained at the left-hand side of the curled arrow to what is contained at the right-hand side of the arrow.

$\Rightarrow$ (Involving). A symbol denoting the *involving* relation between two situation types. Situation theory characterises *constraints* by introducing a primitive relation between types of situations, the relation of *involving*. The involving relation may be introduced on the basis of abstract links that capture (are) systematic regularities (e.g., natural laws, linguistic rules,conventions, logical rules, etc.) connecting situations of one kind with situations of another. With reference to the situation-theoretical framework, a constraint on collaborative design *C* can be expressed by a situation, in which there is a *leading to* relation between two different design states, involving another situation *S'*, by writing:

Constraint *C*: *Design State* $R_1 \rightarrow$ *Design State* $R_2 \Rightarrow S'$

Notes

1 This is true even in a case as acclaimed as the contemporary Italian architect Renzo Piano; in his preface to Piano's recent book, Kenneth Frampton gives a summary as thus (Piano and Brignolo 1997): "In terms of professional politics Piano has steered an independent course, while remaining faithful to the critical ethic of teamwork (perhaps no one practising today has acknowledged his collaborators more fully) and to the inescapable truth that in the field of Baukunst there is no single author."

2 AutoCAD is a trademark of Autodesk, Inc. (http://www.autodesk.com/products/index.htm); MicroStation is a trademark of Bentley Systems Inc. (http://www.bentley.com/products/ mstation/j/); VectorWorks (previously MiniCAD) is a trademark of Diehl Graphsoft Inc. (http://www.diehlgraphsoft.com/); ArchiCAD is a product of GRAPHISOFT (http://www.graphisoft.com/products/); formZ is a product of auto·des·sys, Inc. (http://www.formz.com/); 3D Studio MAX is a trademark of Autodesk, Inc. (http://www.ktx.com/html/products.html).

3 Macromind Director is a product of Macromedia, Inc. (http://www.macromedia.com/software/director/); Adobe Premiere is a product of Adobe Systems Inc. (http://www.adobe.com/products/ premiere/main.html).

4 See CAAD Futures' Web site at http://www.caadfutures.arch.tue.nl/; eCAADe at http://www.ecaade.org/; ACADIA at http://www.acadia.org/home.html; CAADRIA at http://www.caadria.org/; SIGRADI at http://216.167.29.141/sigradi/index.html

5 To some readers, it may be controversial to include the observations and systems done by Gerhard Fischer's group in the survey. However, the inclusion is intended to establish a wider scope of shared drawing activities. In particular, as mentioned in the beginning of the chapter, it is useful to examine any possible links between supporting shared drawing activities and supporting collaborative design. The studies done by Fischer's group present a different research agenda of supporting shared drawing activities involving participants' indirect (i.e., asynchronous) communication in the course of exchanging domain knowledge. It should appear clearer later that graphical construction is an integral part of the overall collaborative design processes the systems developed by Fischer's group, though it may appear highly task-oriented.

6 Similar time space matrices have been proposed by Robert Johansen and others (Johansen et al. 1988) and Clarence Ellis and others (Ellis et al. 1991) in their discussions of groupware design.

7 The argumentation here refers to the argumentation in the Issue Based Information Systems (IBIS) method originally developed by Kunz and Rittel (Kunz and Rittel 1970), and extended by Conklin and Begeman in the gIBIS tool (Conklin and Begeman 1988). According to Fischer et al., the issue-based argumentation method is intended as an interpretation of the "reflection" in the design theory of reflection-in-action proposed by Donald Schön (Schön 1985).

8 As a method parallel to argumentation, critiquing (the generation and sending of critic messages) was developed by Fischer's group to identify and explain why a given design construction is inconsistent with the state of group memory (the so called breakdown situations). The critiquing method has been experimented with by the same research group in implementing several cooperative problem solving systems (see [Fischer and others 1991], for more details). For the theory and practice of expert critiquing systems, a comprehensive survey can be found in (Silverman 1992).

9 For the rest of the book, all illustrations quoted were given permissions either from the original author(s) or the publishers.

10 The concepts and algorithms of flexible coupling and coupling awareness were firstly explored and used in programming multi-user interfaces by Prasun Dewan and Rajiv Choudhary. (See Dewan and Choudhary 1991a; Dewan and Choudhary 1991b for more details.)

11 Note that multiple views of a design are different from multiple views of a drawing. In a kitchen design, for example, multiple views such as structure, lighting, and mechanical services, can be involved, and each view may produce drawings in a distinct domain of construction. Multiple views of, say, a sketch of a kitchen plan, on the other hand, require multiple interpretations of a single graphical construction from different views. For an exposition of a multi-layered model for interpreting architectural drawings, see (Vergopoulos 1991).

12 vmacs is a trademark of the Performing Graphics Company.

13 For example, in the prototype design of Commune, Sara Bly and Scott Minneman described the decision of modelling the system as a "shared pad of paper," which led to the use of a horizontally positioned drawing surface and a writing tool like a pen (Bly 1988).

14 In Patterson's terms (Patterson 1991), the centralised application and conference process is the single abstraction process containing the abstraction objects for the application; and a participant process is a view process containing the view objects for a particular user.

15 More detailed explanations of the difference between the two communication primitives can be read in (Bal et al. 1989).

16 TEXT-GRAPHIC-QUERY is a trademark of the Graphics Performing Company, see Web site at http://www.pgc.com/pgc/home.html.

17 VISUAL-MAIL is a trademark of the Performing Graphics Company.

18 In "Bulletin No. 41 of the International Association of Plastic Arts," UNESCO, Paris 1961, quoted in (Middleton 1967, 278-279).

19 To borrow the example of information flow given in (Barwise and Perry 1983), we may think of a life sketch where Jane has a dog, Mori, who was injured in an accident. Jane later brought Mori to the vet, Fred. Fred took an X-ray picture of Mori and saw a bone fracture in the picture. Jane was then told by Fred that Mori had a fracture in her left leg. The example shows that the information about Mori's broken left leg flows from the situation where Mori was injured to the situation where Jane was aware of the fact that Mori had a broken left leg.

20 Consider the example where a person acquires the age of a (dead) tree by inspecting the tree trunk left on the ground. There are the representation (the tree stump), and the item of information (the tree is, say, 45 years old). It is because he is attuned to the constraint "the age of a tree is equal to the number of rings on the tree stump" that the person acquires that item of information.

21 Also, using tools from situation theory, Devlin and Rosenberg analysed and described how speaker and listener cooperated to achieve shared understanding in the course of communicative, natural language interaction (Devlin 1993); and in (Devlin 1992b), Devlin outlined a working programme showing that situation theory could be a potential framework for the design of interactive information systems.

22 The term "agent" here refers to human authors. An agent as in agency is used to denote that the author or authors of a modelling space may not be the end users of the modelling spaces themselves.

23 Here, a design representation scheme formed by an individual is assumed to be visible to us as observers. In reality, this may not be necessarily so in the sense that the designer has not written or drawn the scheme in any visible form but in his or her head.

24 The notion of a designer's intent in the acts of generating design expressions should be emphasised, because a representation scheme on its own cannot motivate or explain different (specific) expressions even if different persons employ the same representation scheme.
25 Recall the exposition in Chapter 3, I discussed that the underlying graphical structures of joint design expressions can be of two kinds: singular and complex. In the case of complex structures, parts of joint design expressions can be accessed and manipulated in a group modelling space.
26 See, among others, Jeffrey Huang's applying the "Coordination Theory" originally developed by Thomas Malone and Kevin Crowston at the MIT in proposing a systematic collaborative design process (Huang 1999a).
27 This proposition does not cover the possibility that design metaphors may be developed personally from the outset and are then communicated to other members of the design team. That is to say, metaphors or images exist prior to a group meeting not as a consequence of the meeting. In my view, this presents a completely different scenario of collaborative design, which I shall discuss later in Chapters 5 and 6.
28 The World Wide Web is a piece of software invention that has transformed the use of the Internet at a global scale. It is now universally recognised that the British physicist and computer scientist Dr Tim Berners-Lee was the founding father of the Web. The original research proposal that Tim Berners-Lee wrote to persuade CERN to fund his research on developing a Hypertext-based information browser (named as "Mesh") can be read at http://www.w3.org/History/1989/proposal.html. Dr Berners-Lee, now the Director of the World Wide Web Consortium, has charted the history of the Web development since its inception in late 1990 in his recent book (Berners-Lee 1999).
29 For a short history of the Internet, see a well-written Web page posted on the Internet Society's Web site at http://www.isoc.org/internet/history/brief.html.
30 AltaVista is a trademark of AltaVista Company, see Web site at http://www.altavista.com/.
31 Excite is a trademark of Excite Inc., see Web site at http://www.excite.com/.
32 Lycos is a registered trademark of Carnegie Mellon University, see Web site at http://www.lycos.com/.
33 Yahoo! is a trademark of Yahoo Inc., see Web site at http://www.yahoo.com/.
34 The actual construction of Colinia Güell church did not commence until 1908.
35 Again, to use the negotiation situation described earlier, this is to say that B and/or C must find a way to let A know that A's intention in making the change in their shared generic structure is not acceptable.
36 For more information about the Apache's Batik Tool Kit project, interested readers can visit its Web site at http://xml.apache.org/batik/
37 Of course, this does not imply that for a same graphic image the size of an SVG file is always nine times less than that of a JPEG. It can be more or less in other cases. The point is that, with SVG format, software applications can play a greater role in generating graphic images without resorting to binary data sets. From the viewpoint of reducing download time over the Internet, SVG or other SVG-like formats (such as the IVL format developed by ILOG) are beneficial.
38 Currently, two versions of aecXML Schema Specification have been announced: Version 0.81 (September 01, 1999) through the BizTalk submission; and Version 0.87 (September 15, 1999) through the Inaugural aecXML Meeting. Both are available online at http://www.aecxml.org/technical/schema/index.htm.
39 For more latest details of the project, visiting eConstruct Project Web site at http://www.econstruct.org/.
40 The complete example file in gbXML can be accessed online at http://www.idea-server.com/smalloffice.xml.
41 Allegedly, the first Webcam application in the UK was set up at the Computer Laboratory University of

Cambridge for monitoring the supplies of a coffee machine. See the Coffee Machine Web page at http://www.cl.cam.ac.uk/coffee/coffee.html.

42 MASSIVE (Model, Architecture and System for Spatial Interaction in Virtual Environment), see Web site at http://www.crg.cs.nott.ac.uk/research/systems/MASSIVE/.

43 COVEN (Collaborative Virtual Environment), see Web site at http://coven.lancs.ac.uk/home.htm

44 Responsive Workbench, see Web site at http://www-graphics.stanford.edu/projects/RWB/

45 VersaBench™ is a trademark of Fakespace Systems Inc., see Web site at http://www.fakespace.com/products/versa.html.

46 The International Workshop on Cooperative Buildings (CoBuild) has a Web site at http://www.cs.cmu.edu/~CoBuild99/CoBuildHome.html.

47 For updated details, see SmartDesk's Web site at http://www-white.media.mit.edu/vismod/demos/smartdesk/

48 The Dolphin system was developed by the CONCERT group at the GMD-IPSI, for more details about the Dolphin project, see its Web site at http://www.darmstadt.gmd.de/publish/ocean/activities/internal/dolphin.html.

49 Habanero® is a registered trademark owned by The Board of Trustees of the University of Illinois. The Habanero framework and associated applications were developed by the National Center for Supercomputing Applications at the University of Illinois at Urbana-Champaign, for downloading the software and user documents, see their Web site at http://havefun.ncsa.uiuc.edu/habanero/.

50 blaxxun Contact is a product of blaxxun interactive, more details at http://www.blaxxun.com/index.html.

51 Viscape is a product of SUPERSCAPE, see Web site at http://www.superscape.com/index.htm.

References

Ackerman, Mark S. (1998). Augmenting organizational memory: a field study of answer garden. *ACM Transactions on Information Systems* Vol. 16, No. 3. Pp. 203-224.

Ackerman, M. S., and T. W. Malone (1990). Answer Garden: a tool for growing organizational memory. In *Proceedings of the conference on Office information systems*. Cambridge, Mass., USA, April 25-27. Pp. 25-27.

aecXML Domain Committee (2000). aecXML White Paper: A Framework for Electronic Communications for the AEC Industries. The IAI aecXML Domain Committee. (Available online: http://www.aecxml.org/docs/aecwhite.doc)

Agrawala, M., A. C. Beers, I. McDowall, B. Fröhlich, M. Bolas, and P. Hanrahan (1997). The two-user Responsive Workbench: support for collaboration through individual views of a shared space. In *Proceedings of the 24th annual conference on Computer graphics & interactive techniques*. Pp. 327-332.

Aish, R. (1999). Migration from an individual to an enterprise computing model and its implications for AEC Research. *Berkeley-Stanford CE&M Workshop: Defining a Research Agenda for AEC Process/Product Development in 2000 and Beyond.* 26-28 August 1999, Stanford University. (Available on-line: Universityhttp://www.ce.berkeley.edu/~tommelein/CEMworkshop.htm)

Aish, R. (2000a). Computing Fundamentals and Process Re-engineering for Collaborative Design. *MicroStation Manager (MSM) Online*, June 2000, Bentley Systems Inc. (Available on-line: http://www.msmonline.com/jun00/ aish.htm)

Aish, R. (2000b). Custom Objects: a model-oriented end-user programming environment. In *Workshop on Visual Languages for End-User and Domain-Specific Programming*, September 10, 2000, Seattle, WA, USA. (Available on-line: http://www.cs.dal.ca/~smedley/veu/)

Amor, R., and L. Newnham (1999). CAD Interfaces to the ARROW Manufactured Product Server. In *Proceedings of the Eighth International Conference on Computer Aided Architectural Design Futures*. Atlanta, USA, 7-8 June. Pp. 1-11.

Bal, H. E., J. G. Steiner, and A. S. Tanenbaum (1989). Programming languages for distributed computing systems. *ACM Computing Survey* Vol. 21, No. 3. Pp. 261-322.

Barnden, J. A, and K. J. Holyoak (Eds). (1994). *Analogy, Metaphor, and Reminding*. Exeter: Intellect Limited.

Barwise, J. (1989). *The Situation in Logic*. CSLI Lecture Notes No. 17, Stanford, Calif.: Center for the Study of Language and Information.

Barwise, J., and J. Perry (1983). *Situations and Attitudes*. Cambridge, Mass.: MIT Press.

Bentley, K. (1998). ProjectBank: Eliminating the Chaos of Managing Engineering Project Information (White Paper), Bentley Systems Incorporated, October 15, 1998. (Available on-line: http://www.bentley.com/products/projbank/white.htm)

Berners-Lee, T. (1999). *Weaving the Web*. London: Orion Business.

Berners-Lee, T., J. Hendler and O. Lassila (2001). The Semantic Web. *Scientific American*, May 2001, Pp. 29-37.

Bijl, A. (1986). Designing with words and pictures in a logic modelling environment. In *Proceedings of Computer-Aided Architectural Design Futures '85*. Delft, The Netherlands, 18-19 September. Pp. 128-145.

Bijl, A. (1987). Strategies for CAD. In *Intelligent CAD Systems 1: Theoretical and Methodological Aspects*. Edited by P. J. W. ten Hagen and T. Tomiyama. Berlin: Springer-Verlag. Pp. 2-19.

Bijl, A. (1989). *Computer Discipline and Design Practice: Shaping Our Future*. Edinburgh Information Technology Series, edited by S. Michaelson, M. Steedman, and Y. Wilks. Edinburgh: Edinburgh University Press.

Bijl, A., T. Renshaw, and D. F. Barnard (1970). The Use of Graphics in the Development of Computer Aided Design for Two Storey Houses. In *Use of Computers for Environmental Engineering Related to Buildings*. Edited by T. Kusuka: US National Bureau of Standards. Pp. 21-36.

Bijl, A., D. Stone, and D. Rosenthal (1979). Integrated CAAD Systems. Technical Report, Final Report of DoE funded research project DGR 470/12. EdCAAD, Department of Architecture, University of Edinburgh, Edinburgh, Scotland.

Birmingham, W., T. Darr, E. Durfee, A. Ward, and M. Wellman (1993). Supporting mechatronic design via a distributed network of intelligent agents. In *AAAI-93 Workshop on AI in Collaborative Design*. Washington D.C., July 11-15. Pp. 15-34.

Blundell Jones, P. (1992). Holy Vessel. *Architects' Journal*, No. 1 July 1992. Pp. 24-37.

Blundell Jones, P., A. Williams, and J. Lintonbon (1999). The Sheffield Urban Study Project. *Architectural Research Quarterly* Vol. 3, No. 3. Pp. 235-244.Bly, S. L. (1988). A Use of Drawing Surfaces in Different Collaborative Settings. In *Proceedings of the Conference on Computer Supported Cooperative Work (CSCW'88)*. Pp. 250-256.

Boyer, C. M. (1996). *CyberCities*. New York: Princeton Architectural Press.

Böhms, M., G. te Brake, P. Bonsma, F. Tolman and R. van Rees (2001). Draft version of the bcXML Specification, IST-1999-10303, D103draft. (Available online: http://www.econstruct.org/6-Public/ d103draft_v3.doc)

Bray, T., J. Paoli and C. M. Sperberg-McQueen (Eds.) (2000). Extensible Markup Language (XML) 1.0 (Second Edition). W3C Recommendation 6 October 2000. (Available online: http://www.w3.org/ TR/REC-xml#sec-intro)

Brinck, T. , and L. M. Gomez (1992). A Collaborative Medium for the Support of Conversational Props. In *Proceedings of CSCW'92*. Pp. 171-178.

Broadbent, G. (1973). *Design in Architecture: Architecture and the Human Sciences*. New York: John Wiley and Sons.

Bucciarelli, L. (1988). An ethnographic perspective on engineering design. *Design Studies* Vol. 9, No. 3. Pp. 159-168.

Cartwright, D., and Zander A. (1968). *Group Dynamics: Research and Theory*. London: Tavistock Publications.

Chang, S. F., J. R. Smith, M. Beigi, and A. Benitez (1997). Visual Information Retrieval from Large Distributed Online Repositories. *Communications of the ACM* Vol. 40, No. 12. Pp. 63-71.

Chen, N., T. Kvan, J. Wojtowicz, D. Bakergem, T. Casaus, J. Davidson, J. Fargas, K. Hubbell, W. Mitchell, T. Nagakura, and P. Papazian (1994). Place, Time and The Virtual Design Studio. In *ACADIA '95*. Washington University, Saint Louis, USA. Pp. 115-132.

Chiu, M. L. (1995). Collaborative design in CAAD studios: shared ideas, resources, and representations. In *Proceedings of the Sixth International Conference on Computer-Aided Architectural Design Futures (CAAD Futures '95)*. Singapore, September 24-26. Pp. 749-759.

Christel, M., D. Wood, and S. Stevens (1993). AMORE: The Advanced Multimedia Organizer for Requirements Elicitation. Technical Report, CMU/SEI-93-TR-012. Software Engineering Institute, Carnegie Mellon University, Pittsburgh.

Clarke, J. A. (1985). *Energy Simulation in Building Design*. Bristol: Hilger.

Collins, G. R., and J. B. Nonell (1983). *The Designs and Drawings of Antonio Gaudí*, edited . Princeton, N. J., Guildford: Princeton University Press.

Conklin, J., and M. L. Begeman (1988). gIBIS: A Hypertext Tool for Exploratory Policy Discussion. *ACM Transaction on Office Information Systems* Vol. 6, No. 4. Pp. 303-331.

Conforti, F., and R. Sahai (2001). *Inside MicroStation: Updated for MS/J and ProjectBank DGN*. Albany, N.Y.: Delmar Publishers.

Coyne, R. (1995). Computers, metaphors and change. *Architectural Research Quarterly* Vol. 1, No. 1. Pp. 62-67.

Coyne, R., J. Lee, D. Duncan, and S. Ofluoglu (1998). Applying Web-Based Product Libraries. In *Proceedings of EuropIA'98: Cyberdesign*. Paris. Pp. 105-117.

Coyne, R. D., and J. R. Lee (1997). CAD On-line. In *Proceedings of the Sixth International EuropIA Conference*. Edinburgh, Scotland. Pp. 63-75.

Crampton, C. (1987). MUSK: A Multi-User Sketch Program. In *Proceedings of the European UNIX Systems User Group*. Pp. 17-29.

Devlin, K. (1991). *Information and Logic*. Cambridge: Cambridge University Press.

Devlin, K. (1992a). Situation Theory and Social Structure. In *Proceedings of the Conference on Applied Logic*.

Devlin, K. (1992b). Situation Theory and the Design of Interactive Information Systems. Technical Report, CSLI-92-171. Center for the Study of Language and Information, Stanford University, Stanford, Calif.

Devlin, K. and Rosenberg, D. (1993). Situation Theory and Cooperative Action. In *Situation Theory and Its Applications (3)*. Edited by P. Aczel, D. Israel, Y. Katagiri, and S. Peters. Stanford: CSLI Publications.

Dewan, P., and R. Choudhary (1991a). Flexible User Interface Coupling in a Collaborative System. In *Proceedings of the ACM CHI'91 Conference*. Pp. 41-48.

References

Dewan, P., and R. Choudhary (1991b). Primitives for Programming Multi-User Interfaces. In *Proceedings of the ACM Symposium on User Interface Software and Technology (UIST'91)*. Pp. 69-78.

Dretske, F. L. (1981). *Knowledge and the Flow of Information*. Oxford: Blackwell.

Ellis, C. A. , S. J. Gibbs, and G. L. Rein (1991). Groupware: Some Issues and Experiences. *Communications of the ACM* Vol. 34, No. 1. Pp. 39-58.

Fahlén, L. E., O. Ståhl, C. Carlsson, and C. G. Brown (1993). A space based model for user interaction in shared synthetic environments. In *Proceedings of the conference on Human factors in computing systems*. INTERCHI'93, 24-29 April. Pp. 43-48.

Ferraiolo, J. (2000). *Scalable Vector Graphics (SVG) 1.0 Specification Part 1*. New York, N.Y.: iUniverse.Com, Inc.

Fischer, G., J. Grudin, A. Lemke, R. McCall, J. Ostwald, B. Reeves, and F. Shipman (1992). Supporting indirect, collaborative design with integrated knowledge-based design environments. *Human Computer Interaction* Vol. 7, No. 3. Pp. 281-314.

Fischer, G., J. Grudin, R. McCall, J. Ostwald, D. Redmiles, B. Reeves, and F Shipman (1999). Seeding, Evolutionary Growth and Reseeding: The Incremental Development of Collaborative Design Environments. In *Coordination Theory and Collaboration Technology*. Edited by G. M. Olson, T. W. Malone, and John B. Smith. London: Lawrence Erlbaum Associates.

Fischer, G., A. Lemke, T. Mastaglio, and Morch A. I. (1991). The role of critiquing in cooperative problem solving. *ACM Transactions on Information Systems* Vol. 9, No. 3. Pp. 123-151.

GeoPraxis, Inc. (2000). *gbXML XML Schema*. (Available online: http://www.idea-server.com/GreenBuildingXDR.htm)

Gleicher, M., and A. Witkin (1994). Drawing with Constraints. *The Visual Computers* Vol. 11, No. 1. Pp. 39-51.

Glushko, R. L., J. M. Tenenbaum, and B. Meltzer (1999). An XML Framework for Agent-based E-commerce. *Communications of the ACM* Vol. 42, No. 3. Pp. 106-114.

Goguen, J. A., and T. Winkler (1988). Introducing OBJ3. Technical Report, SRI-CSL-88-9. SRI International, Computer Science Laboratory.

Goldschmidt, G. (1988). Interpretation: its role in architectural design. *Design Studies* Vol. 9, No. 4. Pp. 235-245.

Graphisoft (1997). *ArchiCAD for TeamWork User's Guide*. Graphisoft R&D Rt. (Graphisoft Park) Záhony utca 7. 1031 Budapest, Hungary. (Available online http://www.graphisoft.com/support/reference_manuals/docs_actw.html)

Greenberg, S., M. Roseman, D. Webster, and R. Bohnet (1991). Issues and Experiences Designing and Implementing Two Group Drawing Tools. Technical Report, 91/438/22. Department of Computer Science, University of Calgary.

Greenberg, S., M. Roseman, D. Webster, and R. Bohnet (1995). Human and Technical Factors of Distributed Group Drawing Tools. In *Groupware for Real-Time Drawing: A Designer's Guide*, Pp. 37-62, S. Greenberg, S.Hayne, and R. Rada (eds), Maidenhead: McGRAW-HILL Book Companay Europe, 1995.

Greif, I. ed(s). (1988). *Computer-Supported Cooperative Work: A Book of Readings*. San Mateo, Calif.: Morgan Kaufmann.

Gross, M. D., S. M. Ervin, J. A. Anderson, and A. Fleisher (1988). Constraints: Knowledge representation in design. *Design Studies* Vol. 9, No. 3. Pp. 133-143.

Grudin, J. (1991). Introduction to the Special Issue on CSCW. *Communications on the ACM* Vol. 34, No. 12. Pp. 30-34.

Gupta, A., S. Santini, and R. Jain (1997). In Search of Information in Visual Media. *Communications of the ACM* Vol. 40, No. 12. Pp. 35-42.

Habraken, N. J., and M. D. Gross (1988). Concepts design games. *Design Studies* Vol. 9, No. 3. Pp. 150-158.

Hagsand, O. (1996). Interactive Multiuser VEs in the DIVE System. *IEEE MultiMedia* Vol. 3, No. 1.

Halprin, L. (1969). *The RSVP Cycles: Creative Processes in the Human Environment*. New York: George Braziller Inc.

Harold, E. R. and W. Scott Means (2000). *XML in a Nutshell*. Farnham: O'Reilly UK.

Hertzberger, H. (1991). *Lessons for Students in Architecture*. Translated by Ina Rike, Rotterdam: Uitgeverij 010.

Hoskins, E. M. (1977). The OXSYS System. In *Computer Applications in Architecture*. Pp. 343-391.

Huang, J. (1999). Project Extranets and Distributed Design: The Value of Internet-Based Media for Design Collaboration. *ACADIA Quarterly* Vol. 18, No. 3. Pp. 16-18.

Ishii, H., and K. Arita (1991). ClearFace: Translucent Multiuser Interface for TeamWorkStation. In *Proceedings of ECSCW'91*. Pp. 163-174.

Ishii, H., and M. Kobayashi (1992). ClearBoard: A Seamless Medium for Shared Drawing and Conversation with Eye Contact. In *Proceedings on Human Factors in Computing Systems*. Monterey, Calif., USA. Pp. 525-532.

Ishii, H., and N. Miyake (1991). Toward an open shared workspace: computer and video fusion approach of TeamWorkStation. *Communications of the ACM* Vol. 34, No. 12. Pp. 37-50.

Ishii, H., and M. Ohkubo (1990). Design of TeamWorkStation: A Realtime Shared Workspace Fusing Desktops and Computer Screen. In *Proceedings of IFIP WG8.4 Conference on Multi-User Interfaces and Applications*.

Ishii, H., C. Wisneski, S. Brave, A. Dahley, M. Gorbet, B. Ullmer, and P. Yarin (1998). ambientROOM: integrating ambient media with architectural space. In *Proceedings of the CHI 98 summary conference on CHI 98 summary: human factors in computing systems*. Los Angeles. Pp. 173-174.

Janke, R. (1978). *Architectural Models*. London: Academy Editions.

Johansen, R., J. Charles, R. Mittman, and P. Saffo (1988). *Groupware: Computer Support for Business Team*. Series in Communication Technology and Society, edited . New York: The Free Press.

Kahn, E. J. (1935). *Design in Art and Industry*. New York: C. Scribner's Sons.

Kim, H. V. (1997). Frontiers in Electronic Media. *Interactions* Vol. 4, No. 4. Pp. 32-64.

Koblick, R. (1999). Concordia. *Communications of the ACM* Vol. 42, No. 3. Pp. 96-97.

Kolodner, J. L. (1993). *Case-Based Reasoning*. San Mateo, Calif.: Morgan Kaufmann Publishers.

Krishnamurti, R. (1985). A Model for Design Description. *Edinburgh Architecture Research* Vol. 12. Pp. 71-89.

Krüger, W., C. Bohn, B. Frohlich, H. Scheuth, W. Strauss, and G. Wesche (1995). The responsive workbench: A virtual work environment. *IEEE Computer* Vol. 28, No. 7. Pp. 42-48.

Krüger, W., and B. Frohlich (1994). The responsive workbench. *IEEE Computer Graphics and Applications*, No. May. Pp. 12-15.

Kunz, W., and H. W. J. Rittel (1970). Issues as Elements of Information Systems. Technical Report, 131. Institute of Urban and Regional Development, University of California, Berkeley, Calif.

Lakin, F. (1983). Measuring Text-Graphic Activity. In *Proceedings of the GRAPHICS INTERFACE'83*. Edmonton, Alberta.

Lakin, F. (1986). Spatial parsing for Visual Languages. In *Visual Languages*. Edited by S. K. Chang, T. Ichikawa, and P. A. Ligomenides. New York: Plenum Press. Pp. 35-85.

Lakin, F. (1988). A Performing Medium for Working Group Graphics. In *Computer Supported Cooperative Work: A Book of Readings*. Edited by Irene Greif. San Mateo, Calif.: Morgan Kaufmann Publishers. Pp. 366-396.

Lakin, F. (1990). Visual Languages for Cooperation: A Performing Medium Approach to Systems for Cooperative Work. In *Intellectual Teamwork: Social and technological Foundations of Cooperative Work*. Edited by J. Galegher, R. E. Kraut, and C. Egido. Mahwah, N.J.: Lawrence Erlbaum Associates, Inc. Pp. 453-488.

Lakoff, G., and M. Johnson (1980). *Metaphors We Live By*. Chicago: University of Chicago Press.

Lam, W. M. C. (1977). *Perception and Lighting as Formgivers for Architecture*. New York: McGraw-Hill Book Company.

Lange, D. B., and M. Oshima (1999). Seven good reasons for mobile agents. *Communications of the ACM* Vol. 42, No. 3. Pp. 88-89.

Lawson, B. R. (1990). *How Designers Think*. 2nd Edition. London: Butterworth Architecture.

Lawson, B. R. (1994). *Design in Mind*. Oxford: Butterworth Architecture.

Lindsay, M., and M. Grant (1997). Research into the Performance of High Quality Video Conferencing Technology using Optical Networks and its Application to the Teaching of Architecture. In *Proceedings of the Sixth International EuropIA Conference*. Edinburgh, April 2-3. Pp. 203-209.

Lu, I. M. (1992). Supporting Idea Management in a Shared Drawing Tool. Unpublished Master Thesis, Department of Computer Science, University of Toronto.

Lu, I. M., and M. Mantei (1991). Idea Management in a Shared Drawing Tool. In *Proceedings of ECSCW'91*. Pp. 97-112.

Maher, M. L., M. Balachandran, and D. M. Zhang (1995). *Case-Based Reasoning in Design*, edited . Mahwah, NJ: Lawrence Erlbaum Associates, Inc.

Martinell, C. (1979). *Gaudí : his Life, his Theories, his Work*. Barcelona: Barcelona Editorial Blume.

Menzel, C., R. J. Mayer, and L. K. Sanders (1992). Representation, Information Flow, and Model Integration. In *Proceedings of the First International Conference on Enterprise Integration Modeling*. Pp. 131-141.

References

Middleton, M. (1967). *Group Practice in Design*. London: Architectural Press.

Minneman, S. L., and S. A. Bly (1991). Managing á Trois: A Study of a Multi-User Drawing Tool in Distributed Design Work. In *Proceedings of the Conference on Human Factors in Computing Systems*. Pp. 217-224.

Mori, T., and H. Nakagawa (1991). A Formalization of Metaphor Understanding in Situation Semantics. In *Situation Theory and Its Applications (2)*. Edited by J. Barwise, J. M. Gawron, G. Plotkin, and S. Tutiya. Stanford, Calif.: CSLI Publications.

Nardi, B. A. (1993). A Small Matter of Programming: Perspectives on End User Computing. Cambridge, Mass.: The MIT Press.

Newton, R. S. (1999). Special Report on aecXML: An Interview with Keith Bentley. *MicroStation Manager Online*, October 1999. (Available online: http://www.msmonline.com/exclusive/aecxml.htm)

Nwana, H. S., J. Rosenschein, T. Sandholm, C. Sierra, P. Maes, and R. Guttmann (1998). Agent-mediated electronic commerce: issues, challenges and some viewpoints. In *Proceedings of the Second International Conference on Autonomous Agents*. Minneapolis, Minn., USA. Pp. 189-196.

Patterson, J. F. (1991). Comparing the Programming Demands of Single-User and Multi-User Applications. In *Proceedings of the ACM Symposium on User Interface Software and Technology (UIST'91)*. Pp. 87-94.

Peng, C., and P. Blundell Jones (1999). Hypermedia Authoring and Contextual Modeling in Architecture and Urban Design: Collaborative Reconstructing Historical Sheffield. In *Media and Design Process (Proceedings of ACADIA '99)*. Salt Lake City, Utah, USA, October 29-31. Pp. 114-125.

Pentland, A. (1998). Smart rooms, desks, and clothes. In *Proceedings of the third international ACM conference on Assistive technologies*. Marina del Rey Calif., USA, April 15-17. Pp. 1-2.

Piano, R., and R. Brignolo (1997). *The Renzo Piano Log Book*. London: Thames and Hudson Ltd.

Platt, J. C., and A. Barr (1988). Constraint methods for flexible models. In *Proceedings of the 15th annual conference on computer graphics*. Pp. 279-288.

Pressman, A. (1995). *The Fountainhead-ache: The Politics of Architect-Client Relations*. New York: John Wiley & Sons, Inc.

Rapaport, R. (1996). In His Image. *WIRED*, November 1996. Pp. 72-106.

Reeves, B., and F. Shipman (1992). Supporting Communication between Designers with Artifact-Centered Evolving Information Spaces. In *Proceedings of CSCW'92*. Pp. 394-401.

Rehak, D. R. (1985). Interfacing expert systems with design databases in integrated CAD systems. *CAD* Vol. 19, No. 9. Pp. 235-245.

Richens, P., and M. Trinder (1999). Exploiting the Internet to Improve Collaboration between Users and Design Team: The Case of the New Computer Laboratory at the University of Cambridge. In *Proceedings of the Eighth International Conference on Computer Aided Architectural Design Futures*. Atlanta, USA, 7-8 June. Pp. 31-47.

Rowe, P. G. (1987). *Design Thinking*. Cambridge Mass.: The MIT Press.

Sahai, R. S. (2000). Inside MicroStation: Understanding the Structure of ProjectBank. *MSM Online*, January 2000. (Also available on-line: http://www.alphacorporation.com/press/press_typb.asp)

Sandholm, T. (1999). Automated negotiation. *Communications of the ACM* Vol. 42, No. 3. Pp. 84-85.

Saxson, R. (1999).Virtually There. *Building Design*, October 8. Pp. 10.

Schmitt, G. (1990). IBDE, VIKA, ARCHPLAN: Architectures for Design Knowledge Representation, Acquisition and Application. In *Intelligent CAD II*. Edited by H. Yoshikawa and T. Holden. Amsterdam; Oxford: North-Holland.

Schodek, D. L. (1980). *Structures*. London: Prentice-Hall.

Schön, D. (1985). *The Design Studio: An Exploration of its Traditions and Potential*. London: RIBA Publications Ltd.

Schön, D. A. (1991). *The Reflective Practitioner: How Professionals Think in Action*. Aldershot Hants: Avebury.

Shelden, D., S. Bharwani, W. Mitchell, and J. Williams (1995). Requirements for virtual design review. *Architectural Research Quarterly* Vol. 1, No. 2. Pp. 80-89.

Shneiderman, B., and P. Maes (1997). Direct Manipulation vs. Interface Agents. *Interactions* Vol. 4, No. 6. Pp. 42-61.

Silverman, B. G. (1992). Survey of Expert Critiquing Systems: Practical and Theoretical Frontiers. *Communications of the ACM* Vol. 35, No. 4. Pp. 106-127.

Sloman, M., and J. Kramer (1987). *Distributed Systems and Computer Network*, London: Prentice-Hall International (UK) Ltd.

Speidel, M. (1991). *Team Zoo*. Translated by Michael Robinson. London: Thames and Hudson Ltd.

Star, S. L. (1989). The Structure of Ill-Structured Solutions: Boundary Objects and Heterogeneous Distributed Problem Solving. In *Distributed Artificial Intelligence Vol. 2 (Research Notes in Artificial Intelligence)*. Edited by L. Gasser and M. N. Huhns. London: Pitman.

Stefik, M., G. Foster, D. Bobrow, K. Kahn, S. Lanning, and L. Suchman (1987). Beyond the chalkboard: computer support for collaboration and problem solving in meetings. *Communications of the ACM* Vol. 30, No. 1. Pp. 32-47.

Stellman, T. A. and G. V. Krishnan (2000). *Harnessing AutoCAD 2000i*. Albany, N.Y.: Autodesk Press.

Streitz, N. A., J. Geißler, and T. Holmer (1998). Roomware for Cooperative Buildings. In *Cooperative Buildings: Integration Information, Organization, and Architecture*. Edited by N. A. Streitz, S. Konomi, and H. Burkhard. Berlin: Springer-Verlag. Pp. 4-21.

Streitz, N. A., J. Geißler, T. Holmer, S. Konomi, C. Müller-Tomfelde, W. Reischl, P. Rexroth, P. Seitz, and R. Steinmetz (1999). i-LAND: an interactive landscape for creativity and innovation. In *Proceeding of the CHI 99 conference on Human factors in computing systems: the CHI is the limit*. Pittsburgh, Pa., USA, May 15-20. Pp. 120-127.

Szovenyi-Lux, M. (1997). ArchiCAD for Teamwork: A New Concept in CAD Teamworking. In *Proceedings of the 15th eCAADe Conference: Challenges of the Future*. (Available online http://www.tuwien.ac.at/ecaade/pro/szovenyi/szovenyi.htm)

Tang, J. C. (1989). Listing, Drawing, and Gesturing in Design: A Study of the Use of Shared Workspace by Design Teams. Technical Report, SSL-89-3. Palo Alto Research Center, Palo Alto.

Tang, J. C. (1991). Findings from observational studies of collaborative work. *International Journal of Man Machine Studies* Vol. 34.

Tang, J. C., and L. J. Leifer (1988). A Framework for Understanding the Workspace Activity of Design Teams. In *Proceedings of the Conference on Computer Supported Cooperative Work (CSCW'88)*. Pp. 26-28.

Tang, J. C., and S. L. Minneman (1991a). VideoDraw: A video interface for collaborative drawing. *ACM Transactions on Information Systems* Vol. 9, No. 2. Pp. 170-184.

Tang, J. C., and S. L. Minneman (1991b). VideoWhiteboard: Video Shadows to Support Remote Collaboration. In *Proceedings of the ACM SIGCHI Conference on Human Factors in Computing Systems*. Pp. 315-322.

Terzopoulos, D., J. Platt, A. Barr, and K. Fleischer (1987). Elastically deformable models. In *Proceedings of the 14th annual conference on Computer Graphics*. Pp. 205-214.

Tjalve, E., M. Andreasen, and F. Schmidt (1979). *Engineering graphic modelling : a workbook for design engineers*. Translated by G. Pitts. London: Newnes-Butterworth.

Tolman, F., R. van Rees, R. Beheshti, M. Böhms, P. Debras, A. Zarli and R. Steinmann (2000). The bcXML Baseline, IST-1999-10303, D101. (Available online: http://www.econstruct.org/6-Public/d101_v4.doc)

Tweed, C., and A. Bijl (1988). MOLE: A Reasonable Logic for Design. *Edinburgh Architecture Research* Vol. 15. Pp. 106-140.

Tzonis, A., and L. Lefaivre (1987). *Classical Architecture: The Poetics of Order*. Cambridge Mass.: The MIT Press.

Uhrskov, F. (2000). *AutoCAD 2000i News*. Holsted, Denmark: Uhrskov Publishing.

Usdin T. and T. Graham (1998). XML: Not a Silver Bullet, But a Great Pipe Wrench. StandardView Vol.6 , No. 3, September 1998, Pp. 125-132, ACM Press.

Vásquez de Velasco, G., and J. J. Trigo (1997). The Tex-Mex Virtual Design Studio. In *Proceedings of the Sixth International EuropIA Conference*. Edinburgh, April 2-3. Pp. 167-180.

Vergopoulos, S. (1991). A Multi-Layered Model for Interpreting Design Drawings. Unpublished Ph.D. Thesis, Deaprtment of Architecture, University of Edinburgh, Edinburgh, Scotland.

Vosniadon, S., and A. Ortony ed(s). (1989). *Similarity and Analogical Reasoning*. Cambridge: Cambridge University Press.

W3C XML Working Group (1998). Extensible Markup Language (XML) 1.0, W3C Recommendation 10-February-1998 (Available online: http://www.w3.org/TR/1998/REC-xml-19980210)

Ward, T. (1987). Design archetypes from group processes. *Design Studies* Vol. 8, No. 3. Pp. 157-169.

White, R. (1992). FireMOLE: an architecture for reasoning with drawings. *Design Studies* Vol. 13, No. 3. Pp. 320-334.

Wilson, P. (1991). *Computer Supported Cooperative Work: An Introduction*. Oxford: Intellect.

Wisneski, C., H. Ishii, A. Dahley, M. Gorbet, S. Brave, B. Ullmer, and P. Yarin (1998). Ambient Displays: Turning

Architectural Space into an Interface between People and Digital Information. In *Cooperative Buildings: Integration Information, Organization, and Architecture*. Edited by N. A. Streitz, S. Konomi, and H. Burkhard. Berlin: Springer-Verlag. Pp. 22-32.

Witkin, A., K. Fleischer, and A. Barr (1987). Energy constraints on parameterized models. *ACM Computer Graphics* Vol. 21, No. 4. Pp. 225-232.

Wolf, C. G., J. R. Rhyne, and L. K. Briggs (1992). Communication and Information Retrieval with a Pen-based Meeting Support Tool. In *ACM Conference on Computer-Supported Cooperative Work (CSCW'92)*. Toronto, 31 October-4 November. Pp. 322-329.

Wong, D., N. Paciorek, and D. Moore (1999). Java-based Mobile Agents. *Communications of the ACM* Vol. 42, No. 3. Pp. 92-102.

Wood, P. D., M. G. Christel, and S. M. Stevens (1994). A Multimedia Approach to Requirements Capture and Modeling. In *Proceedings of the First International Conference on Requirements Engineering*. Colorado Springs, Colo., USA. Pp. 53-56.

Woods, F., and J. Powell (1987). *Overlay Drafting: A Primer for the Building Design Team*. London: Architectural Press.

Wren, C., A. Azarbayejani, T. Darrell, and A. Pentland (1997). Pfinder: Real-Time Tracking of the Human Body. *IEEE Transactions on Pattern Analysis and Machine Intelligence* Vol. 19, No. 7. Pp. 780-785.

Yang, S. A., D. Robertson, and J. R. Lee (1993). KICS: A knowledge-intensive case-based reasoning system for statutory building regulations and case histories. In *The Fourth International Conference on Artificial Intelligence and Law*. Amsterdam, 15-18 June.

Sources and Illustration Credits

The author wishes to acknowledge the following individuals and organisations for permission to reproduce illustrations and table.

Figures

2.2, 2.3, 2.4, 2.5, 2.6, 2.7, 2.9, 2.11, 2.13, 2.15, 2.16, 6.1 and 6.2:
The ACM Press.

2.8, 2.10:
Saul Greenberg, Department of Computer Science, University of Calgary.

2.12:
Lawrence Erlbaum Associates, Inc.

3.1, 3.2 and 3.3:
George Braziller Inc.

3.4, 3.5:
McGraw-Hill Book Company.

3.6:
Verlag Gerd Hatje, Stuttgart.

3.7, 3.8:
MacCormac Jamieson Prichard Architects.

4.2:
The American Museum of Natural History.

5.2, 5.3, 5.4, 5.5 and 5.6:
Princeton University Press.

7.1:
Graphisoft Inc.

7.2:
Autodesk Inc.

7.5:
Adobe Systems Inc.

8.1:
Software Engineering Institute, Carnegie Mellon University.

8.5, 8.6:
IEEE Computer Society

8.7:
Fakespace Inc.

8.9:
Springer-Verlag.

8.10:
(i-LAND), 8.13 (Dolphin): Integrated Publication and Information Systems Institute (IPSI) of GMD, the German National Research Center for Information Technology.

8.13 (Habanero):
National Center for Supercomputing Applications (NCSA).

Table

7.2:
GeoPraxis, Inc.

Further Acknowledgement

The author thanks the following publishers for permission to reproduce edited versions of Chengzhi Peng's previously published material:

Early Experiments in Supporting Collaborative Drawing and Design, Kluwer Academic Publishers.
Collaborative Design and Discovery of Metaphors, europia Productions.
Flexible Generic Frameworks and Collaborative Design, Elsevier Science Ltd.

Index

Index